*A Student Handbook for
Writing in Biology*

A Student Handbook for Writing in Biology

Karin Knisely
Bucknell University

SINAUER ASSOCIATES, INC.

W. H. FREEMAN AND COMPANY

Editors: Andrew D. Sinauer, Sara Tenney
Production Manager: Christopher Small
Cover Design: Jefferson Johnson
Book Design: Kristy Sprott
Page Layout: The Format Group LLC

Address editorial correspondence to:
Sinauer Associates, Inc., 23 Plumtree Road, Sunderland, Massachusetts 01375
U.S.A.
www.sinauer.com

Address orders to:
VHPS/W. H. Freeman & Co. Order Department, 16365 James Madison
Highway, U.S. Route 15, Gordonsville, VA 22942 U.S.A.
www.whfreeman.com

Orders: 1-888-330-8477

Microsoft Word and Microsoft Excel are registered trademarks of Microsoft
Electronics, Inc. In lieu of appending the trademark symbol to each occurrence,
the author and publisher state that these trademarked product names are used
in an editorial fashion, to the benefit of the trademark owner, and with no
intent to infringe upon the trademarks.

ISBN: 0716766469

Printed in U.S.A.
4 3 2 1

To hardworking biology students and their dedicated instructors

CONTENTS

PREFACE

As the title suggests, the goal of this book is to provide practical advice to students who are learning to write according to the conventions in biology. Writing in biology requires a variety of skills, which can be classified broadly into information research, information technology, and good writing skills. Information research involves locating pertinent sources efficiently and evaluating a source's reliability. Information technology is using a word processor to type and format documents according to prescribed standards, as well as using computer software to plot graphs, rather than drawing them by hand. Good writing is a combination of organization, clarity, brevity, and mechanics.

My experiences with first- and second-year college students indicate that most students do not have all of these skills when they are asked to write their first scientific paper—very often a laboratory report. Students need and want specific instructions on all aspects of the preparation process, but in a concise, user-friendly format. My objective in writing this book was to use the standards of journal publication as a model, to provide the tools students need to meet those standards, and to minimize the time and effort students spend looking for specific information.

My approach to teaching biology students good writing skills relies on the same strategies used in writing classes. These strategies are explained in different parts of the book and include, for students:

- To develop sound reading strategies to extract the basic message and the overall structure of the text (Chapter 3)
- To write for a particular audience and with the appropriate tone (Chapters 3 and 4)
- To learn to paraphrase, not only to avoid plagiarism, but also to improve comprehension (Chapters 3 and 4)
- To learn to self-evaluate first drafts critically (Chapter 5)

- To take an active part in the editing process through peer evaluation (Chapter 5 and Appendix 3). When students evaluate each other's papers, they are reading authentic writing on a topic that is relevant to their discipline. Looking for mistakes in their classmate's papers promotes error awareness in their own work. Peer evaluation also encourages collaboration while interpreting data and, in the long-term, promotes professional collaboration.

- To pay much more attention to the revision process (Chapter 5), which includes the awareness that distance is needed to evaluate work objectively.

- To learn from their mistakes (the "Revision Checklist" and "Laboratory Report Mistakes" key). Students must take the time to review their instructor's comments on their papers and keep an error log of frequent writing problems. This kind of awareness is essential for empowering students to become competent and self-sufficient writers.

- To view feedback from the instructor as an opportunity for professional development. To accomplish this goal, feedback from the instructor must neither be too general to be useful ("improve flow") nor so comprehensive that the student has no need to think about possible revisions.

As instructors, we feel an obligation not only to evaluate students' work fairly and impartially, but also—and more importantly—to provide them with timely, constructive criticism. Without this feedback, students continue to repeat their mistakes, and both students and instructors become caught up in a vicious cycle. Constructive feedback takes time, however, and it is this lack of time that prompted me to develop the "Revision Checklist" and the "Laboratory Report Mistakes" key. These checklists have two functions. First, they alert students to specific content and format requirements and to potential errors, based on my past experience with this student audience. Second, they save instructors time by replacing an explanation of an error with a number in the key. It is then the student's responsibility to consult the key for the number's meaning.

The importance of information technology—word processing and graphing on the computer—cannot be overstated. Appearance is everything, whether students are preparing a laboratory report for an introductory biology class or scientists are presenting their research at an international meeting. When students develop technical competence on the computer, they are investing in their future.

Acknowledgments

I would like to recognize the contributions of three lifelong interests on this book: translation, sports, and biology. Learning to speak German at an early age made translation a natural avocation, which turned into a career in freelance technical translation. I am indebted to the translation profession for setting high standards of quality. It is not a trivial task to produce a translation that accurately expresses both the content and tone of the source text and that follows the conventions familiar to users of the target text. Furthermore, it was through translation that I became proficient at using word processors: customer satisfaction is an excellent motivator for getting format and layout right.

My participation in and love of sports has provided me with countless benefits, including an appreciation for preparation, discipline, and teamwork, and a desire always to do one's best. It is this spirit I try to instill in my students, namely, that individual effort assisted by collaborative effort and continual quality control of every endeavor help us become competent and independent professionals.

Biology is fascinating to me because of its practical applications. Observations about life present us with more than a lifetime of material to study. The more we learn, the more we come to appreciate the complexity of life. Trying to make sense of our observations and communicating our knowledge to others is what makes teaching biology such an exciting profession. Especially rewarding is the enthusiasm we generate among our students when we communicate our knowledge well.

Many people helped to produce this book. In particular, I would like to thank the students in my Cell and Molecular Biology and Organismal Biology laboratories and my colleagues in the Biology Department at Bucknell for their many helpful suggestions on how to improve both the content and organization. Thanks to Randy Wayne at Cornell University, who first encouraged me to try to have this material published. John Byram at W. W. Norton & Co. had faith in the project from the start, commissioned reviews of an early draft of the manuscript, and was ultimately responsible for introducing me to Andy Sinauer, who in turn put me in touch with Sara Tenney of W. H. Freeman. Together they provided encouragement, feedback, and organizational input that resulted in the final manuscript and a published book. Kristy Sprott created an attractive design that incorporates many elements with maximum readability, and Carol Wigg shepherded the book through the editorial and production processes.

I am grateful to Wendy Sera at Baylor University, Elizabeth Bowdan at the University of Massachusetts at Amherst, and Mary Been at Clovis

Community College for reviewing an early version of the manuscript and suggesting improvements. I very much appreciate John Woolsey's insightful comments on the purpose of poster presentations and his expert technical advice on effective poster layout. Thanks to Lynne Waldman for allowing me to use her laboratory report as a student model. Joe Parsons allowed me to use the material on how to read textbooks from the University of Victoria Counselling Services Web site. Jim van Fleet, Science and Technology Librarian, and other staff members at Bucknell's Information Services & Resources (library and computer services) helped me with research questions and computer-related tasks. I learned much about writing from the staff at the Bucknell Writing Center.

Finally, I would like to thank my family and friends for their encouragement and love. I am especially grateful to my parents, Elfriede and Adolph Wegener, for instilling in me a love of languages and giving me a positive outlook on life. My children, Katrina, Carleton, and Brian, are my pride and joy. Finally, I thank my husband, Chuck, whose support, sense of humor, and love sustain me.

KARIN KNISELY
LEWISBURG, PA
DECEMBER 2001

Chapter 1

THE SCIENTIFIC METHOD

Trying to understand natural phenomena is human nature. We are curious about why things happen the way they do, and we expect to be able to understand these events through careful observation and measurement. This is known as the scientific method, and it is the foundation of all knowledge in the biological sciences.

An Introduction to the Scientific Method

The scientific method involves a number of steps:

- Asking questions
- Looking for sources that might help answer the questions
- Developing possible explanations (hypotheses)
- Designing an experiment to test the hypotheses
- Collecting data
- Organizing data to help interpret the results
- Developing possible explanations for the experimental results
- Revising original hypotheses to take into account new findings
- Designing new experiments to test the new hypotheses (or other experiments to provide further support for old hypotheses)
- Sharing findings with other scientists

Ask a Question

As a biology student, you are probably naturally curious about your environment. You wonder about the hows and whys of things you observe. To apply the scientific method to your questions, however, the phenomena of interest must be sufficiently well defined. The parameters that describe the phenomena must be measurable and controllable. For example, let's say that you wonder if plant hormones influence plant height. You might state the question in the following terms:

> Will the addition of gibberellic acid (a plant hormone) cause *Brassica rapa* dwarf mutants to grow taller?

This is a question that could be answered using the scientific method, because the parameters can be controlled and measured. On the other hand, the following question could not be answered easily with the scientific method:

> Will the addition of plant hormones increase a plant's sense of well-being?

In this example, "a sense of well-being" is not something that can be measured or controlled.

Look for Answers to Your Question

There is a good chance that other people have already asked the same question. That means that there is a good chance that you may be able to find the answer to your question, if you know where to look. Your textbook, journal articles, and the Internet (see Harnack and Kleppinger, 2001 for how to determine a Web site's reliability) are all good places to begin finding answers. Curiously, attempts to answer the original question often result in new questions, and unexpected findings lead to new directions in research. By reading other people's work, you may think of a more interesting question, define your question more clearly, or modify your question in some other way.

Turn Your Question into a Hypothesis

As a result of your literature search or conversations with experts, you may now have a tentative answer to your original (or modified) question. Now it is time to develop a hypothesis. A hypothesis is a possible explanation for something you have observed. **You must have information before you can propose a hypothesis!** Without information, your hypothesis is nothing more than an uneducated guess. That is why you must look for possible answers before you can turn your question into a hypothesis.

A useful hypothesis is one that can be tested and either supported or negated. A hypothesis can never be proven right, but the evidence gained from your observations and/or measurements can provide support for the hypothesis. Thus, when scientists write papers, they never say, "The results prove that…" Instead, they write, "The results suggest that…" or "The results provide support for…"

Design an Experiment to Test Your Hypothesis

It doesn't matter whether your hypothesis is ultimately negated. A hypothesis is simply a starting point for designing an experiment to test your explanation. Designing an experiment requires creativity, deductive logic, and a sense for what is practical. Creativity involves brainstorming (possibly with others) to view the subject from many different perspectives. Deductive logic starts with certain premises (your observations and readings) and attempts to draw a novel conclusion. A sense for what is practical is required, because your experimental design will be limited by considerations such as availability of equipment and supplies, cost, time, and so on.

When you design an experiment, pay attention to the following points.

Define the variables. There are three kinds of variables: dependent variables, the independent variable, and controlled variables. **Dependent variables** are variables such as growth, number of seeds, number of body segments, number of offspring, and so on that you can measure or observe. These variables may be altered by the experimental conditions.

The *one* variable that a scientist manipulates in a given experiment is called the **independent variable**. In a different experiment, a different independent variable can be manipulated. It is important to have *only one* independent variable in a given experiment, because otherwise you don't know which factor is affecting the dependent variable(s).

There are many variables that could possibly affect the outcome of an experiment. Thus, it is important to vary only one (the independent variable) and to keep all others constant. The ones that remain constant are called the **controlled variables**. If, for example, you decide to investigate the effect of gibberellic acid on plant growth, variables such as temperature, humidity, age of the plants, day length, amount of fertilizer, watering regime, and so on would all have to be kept constant. These are the controlled variables.

Design the procedure. Determine how you will carry out the experiment. The procedure can be based on articles in scientific journals, suggestions from colleagues, intuition or experience, or simply the desire to try out a new idea. The procedure must include the following components.

- **Control treatment.** This is the condition in which the independent variable is held at a predetermined level or is absent. For example, if you are investigating the effect of gibberellic acid on plant growth, the control treatment is a plant that does not receive any gibberellic acid. Instead, the control plant receives a comparable volume of water.

- **Level of treatment for independent variable.** What concentration of gibberellic acid should be tested? We don't want to choose a concentration that is too low, otherwise we might not see any effect. On the other hand, the concentration should not be so high that it is toxic to the plant. The level of treatment is usually based on previous research (journal articles) or preliminary experiments. The level can even be a range that is appropriate for the biological system or organism to be tested.

- **Replication.** A single result is not statistically valid. The experiment must be repeated several times to see whether similar results are found. The results from several trials may be averaged and analyzed using statistical tests.

Make predictions about the outcome of your experiment. For example, you might predict that if *B. rapa* is treated with 5×10^{-3} mg/mL gibberellic acid, then the plants will grow taller than if no gibberellic acid is applied. If the results of the experiment support the prediction, then the hypothesis is supported. If not, then the hypothesis is negated.

Predictions are an important tool, because they give you a sense of direction. But there is a danger here, too: Do not let your predictions affect your objectivity. **Do not make your results fit your predictions—** instead, modify your hypothesis to fit your results.

What is learned from a negated hypothesis can be just as valuable as what is learned from a "successful" experiment. The subsequent modifications of the hypothesis and the associated experiments help researchers gain assurance that their explanation of a particular phenomenon is valid.

Collect Data

Once you have planned and set up your experiment, you are ready to make observations or measurements about your test subject. Consistency is very important in this regard. If, for example, you are measuring plant height, you must always measure according to the same criteria. If you measure from the rim of the pot to the shoot tip the first time, you can't measure from the soil to the shoot tip the next time without introducing error into the results. Sometimes it is necessary to modify during the course of your experiment how you collect data and *what kinds of data* are important. These modifications may also be included when you design your next experiment.

You should realize that even some of the most elementary questions in biology have taken hundreds of scientists many years to answer. One approach to the problem may have seemed promising at first, but as data are collected, problems with the method or other complications may become apparent. Although the scientific method is indeed methodical, it also requires imagination and creativity. Successful scientists are not discouraged when their initial hypotheses are discredited. Instead, they are already revising their hypotheses in light of recent discoveries and planning their next experiment. You will not usually get instant gratification from applying the scientific method to a question, but you are sure to be rewarded with unexpected findings, increased patience, and a greater appreciation for the complexity of biological phenomena.

How to handle variability and unexpected results.   Variability is a fact of life. If you toss a coin 10 times, you may get 5 heads and 5 tails the first time, 4 heads and 6 tails the second time, and 8 heads and 2 tails the third time. Even though you used the same coin and tossed it the same way, you got different results in different trials.

This same kind of variability is likely to be present in experimental data. Even when you apply the same treatment to different individuals of the same species of plant, the individual plants may respond differently. How do you know if the different results are due to variability among individuals, the imprecision in making measurements, or some other factor?

This is where statistical analysis comes in. Statistics can help you determine how far you can trust the results when you are sampling a subset of a population in which individuals differ. Statistics can help you decide if the different measurements you obtained for replicates are significant or not.

Some common statistical tests and their uses are given in Table 1.1. Consult a good statistics text (see the Bibliography: Samuels and Witner, 1999; Moore, 2000; Utts and Heckard, 2002) for details on the different tests.

TABLE 1.1   Common statistical tests and their uses

TEST	APPLICATION
Chi square	To compare how closely the observed or measured data compare to the expected results (e.g., for crosses in genetics)
t-test	To compare the means of two groups
ANOVA	To compare the means of three or more groups

What if your results cannot be explained by variability? Although you would like to give yourself the benefit of the doubt, human error is a possibility, especially in introductory biology laboratory exercises. Human error includes failure to follow the procedure, failure to use the equipment properly, failure to prepare solutions correctly, variability when multiple lab partners measure the same thing, and simple arithmetic errors. If you suspect that human error may have influenced your results, it is important to acknowledge its contribution. If you had time to repeat the experiment, you would first try to eliminate sources of human error, rather than revise your original hypothesis.

Organize the Data

Raw data must be organized before you can begin any interpretation. Two ways to organize data are in tables and figures. Tables have rows and columns and are useful for displaying several dependent variables at the same time or when you want to emphasize the numbers themselves rather than the trend shown by the numbers. Figures are graphs, pictures, diagrams, gel photos, X-ray images, and microscope images—any graphic image that is not a table.

Line graphs and bar graphs are commonly used in data analysis. Line graphs show the effect of the independent variable (the one the scientist manipulates) on a selected dependent variable (the one that changes in response to the independent variable). By convention, the independent variable is plotted on the x-axis, and the dependent variable is plotted on the y-axis. Let's say you treated dwarf B. rapa plants with different gibberellic acid concentrations to see how concentration affects growth over a 3-week period. In that case, time is plotted on the x-axis, and stem height on the y-axis for each group of plants (each line represents one concentration of gibberellic acid).

Bar graphs allow you to compare individual sets of data when the data are non-numerical or discontinuous. In the previous example, if you wanted to compare the *final height* of the plants treated with different concentrations of gibberellic acid, each bar would represent the height of one group of plants as measured at the end of the 3-week period. The data are discontinuous, because different treatment groups are being compared.

The best way to display the data depends on the point you are trying to make. For example, if you want to show that the rate of growth is faster in plants treated with gibberellic acid than that in untreated plants, then a line graph would work well. If you want to emphasize that the end effect of treating plants with gibberellic acid is a taller plant, then a bar graph would be suitable. If you want to compare the actual heights of your plants with those in the literature, then a table would be useful. Also keep in mind that a "picture is worth a thousand words," especially when you are comparing physical differences like height, color, and overall appearance of the organisms. Photographs may call your attention to variables that you didn't measure, but that you might want to consider in future experiments. Photographs are also very effective in poster presentations.

Regardless of how you choose to display the data, you must be honest. Do not exaggerate axes or trends to support your hypothesis, when, in truth, the data do not support it. Show variability when necessary. Use statistical methods to reduce huge amounts of data, and be prepared to explain your reasoning.

Keep in mind that there may be *no difference* between the control and the experimental treatments. If there was no difference, say so, and then try to develop possible explanations for these results.

Try to Explain the Results

Once you have organized the data, you are ready to develop possible explanations for the results. You already found sources on the topic when you developed your hypothesis. Return to this material to try to explain your results. Do your results agree with the findings of other researchers? Do you agree with their explanations? If your results do not agree, try to determine why not. Were different methods, organisms, or conditions employed? What were some possible sources of error?

Revise Original Hypotheses to Take New Findings into Account

If the data support the hypothesis, then you should suggest additional experiments to strengthen the hypothesis. If the data do not support the hypothesis, then you should look for "human error" factors first. If these kinds of factors can be ruled out, suggest modifications to the hypothesis. You may also describe ways to test the new hypothesis. Ideally, scientists will thoroughly investigate a question until they are satisfied that they can explain the phenomenon of interest.

Share Findings with Other Scientists

The final phase of the scientific method is communicating your results to other scientists, either at scientific meetings or through a publication in a journal. When you submit your paper to refereed journals, it is read critically by other scientists in your field, and your methods, results, and conclusions are scrutinized. If any errors are discovered, they are corrected before your results are communicated to the scientific community at large.

Poster sessions are also an excellent way to share preliminary findings with your colleagues. The emphasis in poster presentations is on the methods and the results. The informal atmosphere promotes the exchange of ideas among scientists with common interests. See Chapter 7 for ideas and guidance on how to prepare a poster.

Chapter 2

How to Find Primary References

The development of library research skills is an essential part of your training as a biology student. A vast body of literature is available on just about every topic. Finding exactly what you need is the hard part.

In biology, sources are divided broadly into primary and secondary references. **Primary references** are the journal articles, dissertations, technical reports, or conference papers in which a scientist describes his or her original work. Primary references are written for fellow scientists, in other words, for a specialized audience. The objective of a primary reference is to present the essence of a scientist's work in a way that permits readers to duplicate the work for their own purposes and to refute or build on that work.

Secondary references include encyclopedias, textbooks, and articles in popular magazines. Secondary references are based on primary references, but they address a wider, less-specialized audience. In secondary references, there is less emphasis on the methodology and presentation of data. Results and their implications are described in general terms for the benefit of nonspecialist readers.

You will delve into the biological literature when you write laboratory reports, research papers, and other assignments. Although secondary references provide a good starting point for your work, it is important to be able to locate the primary sources on which the secondary sources are based. Only the primary literature provides you with a description of the methodology and the actual experimental results. With this information, you can draw your own conclusions from the author's data.

Although initially it is difficult to read primary literature, it becomes easier with practice, and your persistence will be rewarded with improved critical thinking skills. A further benefit of reading the primary literature is getting to know the scientists who work in a particular subdiscipline. You may discover that you are sufficiently interested in a subdiscipline to pursue graduate work with one or more of the authors of a journal article. Networking is as important in biology as in other fields.

How do you find primary references that are directly relevant to your topic? The fastest and easiest way is to search computer databases using keywords. Most university libraries have licensing agreements with the companies that produce databases commonly used in biology. These licenses are almost always restrictive, making them available only to faculty, staff, and on-campus students. Thus, if you wish to research a topic using databases such as Science Citation Index, FirstSearch, and ArticleFinder, you must be affiliated with a university or other organization that has such a licensing agreement.

Most of this chapter describes how to find primary references using databases. If you do not have access to these databases, however, you can still locate references the old-fashioned way. This method involves building a bibliography from sources cited in textbooks, journal articles, and other literature (see "References Cited in 'Hits,'" p. 12). This method is laborious and time-consuming because you are doing the physical work of the database, but the end result is often the same.

A disadvantage of using databases is that some older, seminal papers may not be indexed. If your assignment requires a thorough search of the literature, you will have to locate these older references the old-fashioned way.

Current Journal Articles

To find current journal articles in biology, use one of the following databases as a starting point (listed in order of usefulness for introductory biology courses):

- Science Citation Index
- BasicBIOSIS (FirstSearch)
- ArticleFinder.

Science Citation Index is perhaps the most useful of the three databases, because it not only allows you to enter keywords to search by subject, author, and place (either institutional affiliation or geographical location), but it also lists the references cited in the hits. This makes it possible to build a large bibliography on a specific topic very quickly; to search the literature both forward and backward in time once you have found a key hit; and to save time searching, because the references cited in the key hit are likely to be exactly on target for your topic. Instructions on using the three databases are given in Boxes A, B, and C.

Electronic journals. Many scientific journals are now available electronically. Ask the librarian if your library subscribes to any.

Box A. To Use the Science Citation Index

1. Ask your librarian for the Science Citation Index URL (*Uniform Resource Locator*). The URL begins with <http://www.> and defines a unique address on the World Wide Web.

2. Click LOG ON to Web of Science.

3. Click EASY SEARCH.

4. Select checkbox for Science Citation Index Expanded (SCI-EXPANDED)—1981 to present.

5. Click TOPIC search.

6. Type in a keyword(s) for your topic. Be as specific as possible to narrow down the number of hits.

7. Click the title of the hits that sound promising. This opens a window containing the abstract of the article and further information on the publication.

8. At the top of the abstract page under the reference, select Cited References. This option shows you the references that were cited in this article.

9. On the Cited References page, the authors are listed alphabetically, and other useful information such as journal, volume, first page of article, and year of each author's publication is also provided. Click any entry in blue to see the abstract of the cited reference.

Books

Books may be primary or secondary references. Although books such as *Annual Reviews* are considered secondary references, the Literature Cited sections in review articles often are an excellent source of primary references.

The following steps are usually the ones to take when looking for books on a specific topic in an academic library.

1. Look for the Online Catalog on the library's home page.

2. Select Search.

3. Enter keyword(s). Be as specific as possible.

4. Book titles will be listed. Click on any titles that sound promising.

5. Write down the call number and note the availability of the book.

Box B. To Use BasicBIOSIS (FirstSearch)

1. Ask your librarian for the BasicBIOSIS URL.

2. Enter a keyword or keywords separated by the word *and*. The keyword(s) should be as specific as possible to avoid getting hundreds of inappropriate references.

3. Article titles will be listed. Click a title that sounds promising.

4. Read the abstract to see if the article pertains to your topic.
   a. If the article sounds promising, print out the abstract and obtain the article in your library. If your library doesn't have the article, you may be able to order the article through interlibrary loan. (This may take 10 or more days.) Ask your librarian how to do this.
   b. If the article is not appropriate, read the next abstract. If necessary, change the keyword(s) and initiate a new search.

To Use Other Databases in FirstSearch

Depending on your topic, you may find some of the other FirstSearch databases helpful:

- MEDLINE—medicine and the health sciences
- Agriculture—life sciences
- General Science—general science, geology, and earth science

Avoid databases that specialize in science magazine articles, because these articles are not considered to be primary references. Enter a keyword and examine the abstracts as before.

---

6. When you look for the book on the shelf, make sure you browse the titles of other books in the vicinity. Because the Library of Congress cataloging system groups books according to topic, you can often find additional sources shelved nearby.

References Cited in "Hits"

Once you have located a highly relevant journal article, you should browse through the References (Literature Cited) at the end of the article for other relevant journal articles. This kind of search is a very effec-

---

Box C. To Use ArticleFinder

1. Enter keyword(s) in the entry box. The keywords should be as specific as possible to avoid getting hundreds of inappropriate references.
2. The article references will be listed. Click the abstract that sounds promising.
3. Read the abstract to see if the article pertains to your topic.
   a. If the article sounds promising, print out the abstract and obtain the article in your library. If your library doesn't have the article, you may be able to order the article through interlibrary loan. (This may take 10 or more days.) Ask your librarian how to do this.
   b. If the article is not appropriate, read the next abstract. You can also select "View other articles linked to these subjects" (at the end of the abstract window), although many of these links will not pertain to your specific keyword. If necessary, go back and change the keyword(s) and initiate a new search.

---

tive way to expand your bibliography, because it provides specific information on the topic of interest.

The Internet

The Internet connects computers around the world, allowing you to communicate with other computer users on the network through e-mail, chat rooms, and the World Wide Web (WWW, or "the Web").

Connecting to the Internet

To get on the Internet, you need a computer, a modem (to connect your computer to a phone line), an Internet service provider (ISP), and browsing software. If you are affiliated with a school, college, or library, the ISP is probably your organization's computer center; if you are making the connection from your home, you need a commercial ISP such as Prodigy or America Online.

Once you have set up an account with an ISP, you can use a browser to find information on the Internet. Sometimes your browser requires plug-ins (additional software) to work efficiently. Instructions for downloading and installing plug-ins are provided in the browser's message windows.

General Navigation

Figure 2.1 shows the menu bar and the toolbars for one of the two most popular graphic browsers, Netscape Communicator (the other one is Microsoft Internet Explorer). There are navigation buttons, menus, and toolbars available to search for information on the Internet. Hold the cursor over a button to open a pop-up menu that describes the button's function.

To locate information on a specific topic, click the Search button.

Some useful basic commands for navigating Web pages are described in Figure 2.1 A comprehensive explanation can be found in Harnack and Kleppinger (2001).

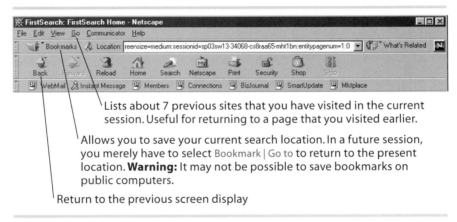

Lists about 7 previous sites that you have visited in the current session. Useful for returning to a page that you visited earlier.

Allows you to save your current search location. In a future session, you merely have to select Bookmark | Go to to return to the present location. **Warning:** It may not be possible to save bookmarks on public computers.

Return to the previous screen display

Figure 2.1 Screen display from Netscape Communicator, with description of some of the useful navigation features

Search Tools

Search tools (also called search engines) enable you to find information on specific topics by typing in keywords. The search tools scan millions of Internet documents for these keywords and then display a list of the documents (with links) in which these keywords appear.

Two of the most useful search tools are AltaVista (<http://www.altavista.com>) and Google (<http://www.google.com>). Both index millions of Web pages and rank the "hits" according to the frequency of occurrence and relevance to the keywords. Metaengines such as Metacrawler (<http://www.go2net.com>) and Ask Jeeves! (<http://ask.com>) are also useful because they allow you to search several Web indexes simultaneously.

For more information on using search tools to find references, see Harnack and Kleppinger (2001).

Evaluate Internet Sources Critically

Although there is an incredible amount of information available on the Internet, much of this information may be unreliable. Whereas journal articles and books have undergone a rigorous review process, most of the information posted on the WWW has not been checked by any authority other than the owner of the Web site.

Be particularly wary of companies and organizations who are more interested in trying to sell a product or an idea than in presenting factual information. The ending of the URL address may help you identify the sponsor, as shown in Table 2.1.

TABLE 2.1 Identifying sponsors of Web sites

TYPE OF WEB PAGE	PURPOSE	ENDING OF URL ADDRESS	EXAMPLES
Informational	To present factual information	.edu, .gov	Dictionaries, directories, information about a topic
Business/ marketing	To sell a product	.com	Coca-Cola, Leica
Advocacy	To influence public opinion	.org	Democratic Party, Republican Party
News	To present very current information	.com	CNN, *USA Today*
Personal	To present information about an individual	Variety of endings, but has tilde (~) embedded in the URL	

Source: Information Services and Resources, Bucknell University, 2001.

Web sites are not considered primary references. When you write a laboratory report or research paper, make sure most of your references are part of the body of published literature, primarily journal articles. You may, however, supplement the literature cited with information retrieved from the Internet.

Chapter 3

READING AND WRITING SCIENTIFIC PAPERS

No matter whether you are a student or already engaged in a profession, writing is a fact of life. There are many reasons for writing: to express your feelings, to entertain, to communicate information, and to persuade. When you write scientific papers, your primary reasons for writing are to communicate information and to persuade others of the validity of your findings.

Types of Scientific Writing

Scientific writing takes many forms. As an undergraduate biology major, you will be asked to write laboratory reports, answer essay questions on exams, write summaries of journal articles, and do literature surveys on topics of interest. Upperclass students may write a research proposal for honors work, and then complete their project by submitting an honors thesis. Graduate students typically write master's theses and doctoral dissertations and defend their written work with oral presentations. Professors write lectures, letters of recommendation for students, grant proposals, reviews of articles submitted for publication to scientific journals by their colleagues, and evaluations of grant proposals. In business and industry, scientific writing may take the form of progress reports, product descriptions, operating manuals, and sales and marketing material.

Hallmarks of Scientific Writing

What distinguishes scientific writing from other kinds of writing? One difference is the motive. Scientific writing aims to inform rather than to entertain the reader. The reader is typically a fellow scientist who intends to use this information, for example, to learn more about a process or to improve a product.

A second difference is the style. Brevity, a standard format, and proper use of grammar and punctuation are the hallmarks of well-written scientific papers. The authors have something important to communicate, and they want to make sure that others understand the significance of their work. Flowery language and "stream of consciousness" prose are not appropriate in scientific writing because they can obscure the writer's intended meaning.

A third difference between scientific and other types of writing is the tone. Scientific writing is factual and objective. The writer presents information without emotion and without editorializing.

Scientific Paper Format

Scientific papers are descriptions of how the scientific method was used to study a problem. They follow a standard format that allows the reader, first, to determine initial interest in the paper, second, to read a summary of the paper to learn more, and, finally, to read the paper itself for all the details. This format is very convenient, because it allows busy people to scan volumes of information in a relatively short time, then spend more time reading only those papers that truly provide the information they need.

Almost all scientific papers are organized as follows:

- Title
- List of Authors
- Abstract
- Introduction
- Materials and Methods
- Results
- Discussion
- References

The Title is a **short, informative description of the essence of the paper**. It should contain the fewest number of words that accurately convey the content. Readers use the title to determine their initial interest in the paper.

Only the names of **people who played an active role** in designing the experiment, carrying it out, and analyzing the data appear in the List of Authors.

The Abstract is a **summary of the entire paper** in 250 words or less. It contains (1) an introduction (scope and purpose), (2) a short description of the methods, (3) results, and (4) conclusions. There are no literature citations or references to figures in the Abstract.

The Introduction concisely states what motivated the study, how it fits into the existing body of knowledge, and the objectives of the work. The Introduction consists of two primary parts:

1. **Background or historical perspective on the topic.** Primary journal articles and review articles, rather than secondary sources such as textbooks and newspaper articles, are cited to provide the reader with direct access to the original work. Inconsistencies, unanswered questions, or new questions that resulted from previous work set the stage for the present study.

2. **Statement of objectives of the work.** What were the goals of the present study?

The Materials and Methods section describes, in full sentences and well-developed paragraphs, **how the experiment was done**. The author provides sufficient detail to allow another scientist to repeat the experiment. Volume, mass, concentration, growth conditions, temperature, pH, type of microscopy, statistical analyses, and sampling techniques are critical pieces of information that must be included. When and where the work was carried out is important if the study was done in the field (in nature), but is not included if the study was done in a laboratory. Conventional labware and laboratory techniques that are common knowledge (familiar to the audience) are not explained. In some instances, it is appropriate to use references to describe methods.

The Results section is **where the findings of the experiment are summarized**, without giving any explanations as to their significance (the "whys" are reserved for the Discussion section). A good Results section has two components:

- A *text*, which forms the body of this section
- Some form of *graphic illustration* that helps the reader visualize the data and get the message faster than from reading a lengthy description

In the Discussion section, the **results are interpreted** and possible explanations are given. The author may:

- Summarize the results in a way that supports the conclusions
- Describe how the results relate to existing knowledge (literature sources)
- Describe inconsistencies in the data. This is preferable to concealing an anomalous result.
- Discuss sources of error
- Describe future extensions of the current work

References list the **outside sources** the authors consulted in preparing the paper. No one has time to return to a state of zero knowledge and rediscover known mechanisms and relationships. That is why scientists rely so heavily on information published by their colleagues. References are typically cited in the Introduction and Discussion sections of a scientific paper, and the procedures given in Materials and Methods are often modifications of those in previous work.

Styles for Citing References

The Council of Biology Editors manual, *Scientific Style and Format* (6th ed.), recommends the following two formats for citing references:

Citation-Sequence System. In the *text*, the citation is given by a number in square brackets or parentheses following the cited sentence. On the *references pages* that follow the Discussion section, the citations are listed in **numerical order** and include the full reference.

Name-Year System. In the *text*, the citation is given in the form of author(s) and year in parentheses following the cited sentence. On the *references pages* that follow the Discussion section, the references are listed in **alphabetical order** according to the first author's last name.

The name-year system has the advantage that people working in the field will know the literature and, on seeing the authors' names, will understand the reference without having to check the reference list. It is more commonly used and generally preferred. With the citation-sequence system, for each reference the reader must turn to the reference list at the end of the paper to gain the same information.

Strategies for Reading Journal Articles

Papers in scientific journals are written by experts in the field. Because you are not yet an expert, you will probably find it difficult to read and understand journal articles. The following strategy may help.

Determine the topic. First, try to determine the topic of the article by reading the title and the abstract. Are the authors trying to answer a specific question, explain observations, present a theoretical model of a process, determine the relationship between one or more variables, or accomplish something else?

Acquire background information on the topic. Read about the topic in your textbook. Because textbook authors generally write for a student audience, not a group of experts, your textbook is likely to be easier to read. See "Strategies for Reading your Textbook" for some ways to read biology textbooks efficiently.

Read the Introduction. The introduction is usually easier to follow than the abstract. Skim the introduction with the following questions in mind:
- Why did the author(s) carry out this work?
- What are the main hypotheses?
- What was previously known about the topic or problem?
- What are the objectives of the current work?

Read the Results section selectively. Look at the figures and tables to determine what variables were studied. The independent variable (the one the investigator manipulated) is plotted on the x-axis, and the dependent variable(s) (the one that changes depending on the independent variable) is plotted on the y-axis. Also look for variables in column headings of tables.

There are two places to look for a qualitative description of each figure and table: the figure/table caption, and in the body of the Results section (text). The caption states the main idea of the graphic illustration. The topic sentence of the paragraph in the text does the same. Subsequent sentences in the paragraph provide details on what trends or findings the reader should notice in each graphic. When you read about the results, ask yourself the following questions:
- What were the independent, dependent, and controlled variables?
- Was there a difference between the controls and the experimental groups?
- What were the main findings regarding the independent and dependent variables?

If necessary, reread the introduction to recall the main objectives and hypotheses of the work. Try to understand the big picture before concerning yourself with the details.

Read the Discussion section. The author typically presents his/her conclusions in this section and describes how the results of the study support these conclusions. This is where you can find out what was learned from the work. In particular:
- Were the hypotheses supported?

- What were the important findings?
- Were there any surprises?
- What further work is necessary or already in progress?
- How does this paper relate to your own work?

Skim the Materials and Methods section. Scan the subheadings (if present) and the topic sentence of each paragraph to identify the basic approach. Do not be concerned with the details at this stage.

Read the article several times. Even experts must read journal articles several times before they understand the methodology and the implications of the findings. Take notes the first time you read the article, noting what is confusing. Consult your notes when you read the article again, and try to clarify what you didn't understand the first time through. Each time you read the article, you will understand a little more.

Strategies for Reading your Textbook

The following strategies are based on the proposition that you cannot read a chapter in a biology textbook just once and understand it completely. Repetition is a key ingredient in learning the material. Repetition not only provides you with multiple opportunities to be exposed to the material, but also gives you time to digest it. The basic approach is to read for organization and key concepts first, and then to fill in the details with each subsequent reading.

The two strategies described here work best with a chapter or section of text no longer than 25–30 pages. The first strategy is proposed by Counselling Services at the University of Victoria, Canada (Palmer-Stone, 2001).

1. Take no more than 25 minutes to:
   - Read the chapter title, introduction, and summary (at the end of the chapter, if present)
   - Read the headings and subheadings
   - Read the chapter title, introduction, summary, headings, and subheadings again
   - Skim the topic sentence of each paragraph (usually the first or second sentence)
   - Skim italicized or boldfaced words
2. Close your textbook. Take a full 30 minutes to:
   - Write down everything you can remember about what you read in the chapter (make a "mind map"). Each time you

come to a dead end, use memory techniques such as associating ideas from your reading to lecture notes or other life experiences; visualizing pages, pictures, or graphs; staring out the window to daydream; letting your mind go blank.

- Figure out how all this material is related. Organize it according to what makes sense in your mind, not necessarily according to how it is organized in the textbook. Write down questions and possible contradictions to check on later.

3. Open your textbook. Fill in the blanks in your mind map with a different colored pencil.

4. Read the chapter again, this time normally. Make another mind map.

A second strategy is:

1. Skim the chapter title, headings, and subheadings for an overview of the chapter content. Write down the headings and subheadings in the form of an outline.

2. Look at your outline and ask yourself the following questions:
   - What is the main topic of this chapter?
   - How do each of the headings relate to the topic?
   - How does each subheading relate to its heading?

3. Read each section, paying special attention to the topic sentence of each paragraph. At the end of each section, summarize the content in your own words. Answer the following questions:
   - What's the point?
   - What do I understand?
   - What is confusing?

4. If you read the assigned pages before the lecture, you can pay attention to the lecture content instead of just frantically taking notes. Check to see if there is a lecture notebook that accompanies your textbook. The lecture notebook contains the figures in black and white, and allows you to take notes during lecture directly on the figures.

5. After the lecture, while the information is still fresh in your mind, reread your notes on your reading. Ask yourself
   - What topics did the instructor emphasize in lecture? Fill in your lecture notes with details from your textbook.
   - What material do I understand better now?
   - What questions remain?

6. Remember that each time you read the material, you will learn a little more.

Study Groups

If you have read the material several times, taken notes, and listened attentively in lecture, but still have questions, talk about the material with your classmates. Small study groups are one reason why students who choose to major in the sciences persist in the sciences, rather than switching to a non-science major (Light, 2001).

What are some benefits of participating in small study groups? One benefit is the comfort level. You may be more likely to talk about problems when you are among your peers; after all, they are not the ones who assign your grade. Secondly, when a group is composed of peers with a similar knowledge base, group members speak the same language. Your instructor speaks a different language, because he or she has already struggled to master the material. When you communicate with your classmates, you verbalize your ideas at a level that is appropriate for your audience of peers. Finally, collaborative learning reflects the way scientists exchange information and share findings in the real world. A spirit of camaraderie develops when people work together toward a common goal. The prospect of learning difficult subject matter is no longer so daunting when you have support from a small group of like-minded individuals. The hard work may even be fun when there is good group chemistry.

Group study is not a substitute for studying alone, however. You must hold yourself accountable for reading the material, taking notes, and figuring out what you do not understand before you meet with your group. If you have not struggled to understand the material yourself, you are not in a position to help a classmate.

Avoid Plagiarism: Paraphrase What You Read

One of the best ways to know whether or not you understand an author's work is to summarize the key findings in your own words. Do not make the mistake of simply copying the text word for word from the textbook or journal article, just because you think that you could not have said it better yourself. Get into the habit of paraphrasing the information in the source document, and carefully noting the source (see "Citation Format" and "Reference Format" in Chapter 4). By convention, quotation marks are not used in scientific writing.

Plagiarism is taking someone else's ideas and passing them off as your own. This includes citing the source but still copying the words verbatim. Usually plagiarism is unintentional and is the result of not having a clear understanding of the material. If you can state in your own words what you think the author meant, then you probably understood it. If not, you may have to read the paper a few more times or ask for help from your instructor or a fellow student.

The Benefits of Learning to Write Scientific Papers

Why is it valuable to learn how to write scientific papers? First, scientific writing is a systematic approach to describing a problem. By writing what you know (and what you do not know) about the problem, it is often possible to identify gaps in your own knowledge.

Secondly, scientific writing aims to persuade the reader of the validity of the procedures, results, and conclusions described in the paper. Improving your reasoning abilities in laboratory reports may have a positive effect on other areas of your life as well.

Third, when you learn to write lab reports, you are investing in your future. Publications in the sciences are affirmation from your colleagues that your work has merit; you have been accepted into the community of experts in your field. Even if your career path is not in the sciences, scientific writing is very logical and organized, characteristics appreciated by busy people everywhere.

Credibility and Reputation

The credibility and reputation of scientists are established primarily by their ability to communicate effectively through their written reports. Poorly written papers, regardless of the importance of the content, may not get published if the reviewers do not understand what the writer intended to say.

You should think about your reputation even as a student. When you write your laboratory reports in an accepted, concise, and accurate manner, your instructor knows that you are serious about your work. Your instructor appreciates not only the time and effort required to understand the subject matter, but also your willingness to write according to the standards of the profession.

Model Papers

Before writing your first laboratory report, go to the library and take a look at some biology journals such as *American Journal of Botany, Ecology, The EMBO Journal, Journal of Biological Chemistry, Journal of Molecular Biology,* and *Marine Biology.* Photocopy one or two journal articles that interest you so you can refer to them for format questions.

Almost all journals devote one page or more to "Instructions to Authors," in which specific information is conveyed regarding length of the manuscript, general format, figures, conventions, references, and so on. Skim this section to get an idea of what journal editors expect from scientists who wish to have their work published.

Because most beginning biology students find journal articles hard to read, a sample student laboratory report is given in Chapter 6. Read the comments in the left margin as you peruse the report to familiarize yourself with the basics of scientific paper format and content, as well as purpose, audience, and tone.

Chapter 4

STEP-BY-STEP INSTRUCTIONS FOR PREPARING A LABORATORY REPORT OR SCIENTIFIC PAPER

In order to prepare a well-written laboratory report or scientific paper according to accepted conventions, the following skills are required:

- A solid command of the English language
- An understanding of the scientific method
- An understanding of scientific concepts and terminology
- Advanced word processing skills
- Knowledge of computer graphing software
- The ability to read and evaluate journal articles
- The ability to search the primary literature efficiently
- The ability to evaluate the reliability of Internet sources.

If you are a first- or second-year college student, it is unlikely that you possess all of these skills when you are asked to write your first laboratory report. Don't worry. The instructions in this chapter will guide you through the steps involved in preparing the first draft of a laboratory report. Revision is addressed in the next chapter, and the Appendices will help you with word processing and graphing tasks.

Timetable

Preparing a laboratory report or scientific paper is hard work. It will take much more time than you expect. Writing the first draft is only the first step. You must also allow time for proofreading and revision. If you work on your paper in stages, the final product will be much better than if you try to do everything at the last minute.

TABLE 4.1 Timetable for writing your laboratory report

TIME FRAME	ACTIVITY	RATIONALE
Day 1	Complete laboratory exercise.	It's fun. Besides, you need data to write about.
Days 2–3	Write first draft of laboratory report.	The lab is still fresh in your mind. You also need time to complete the subsequent tasks before the due date.
Day 4	Proofread and revise first draft (hard copy).	Always take a break after writing the first draft and before revising it. This "distance" gives you objectivity to read your paper critically.
Day 5	Give first draft to a classmate for review.	Your peer reviewer is a sounding board for your writing. He/she will give you feedback on whether what you intended to write actually comes across to the reader. You may wish to answer the questions on the "Draft Self-Assessment" form (see Appendix 3) to alert your peer reviewer to concerns you have about your paper.
	Arrange to meet with your classmate after he/she has had time to review your paper ("writing conference").	An informal discussion is useful for providing immediate exchange of ideas and concerns.

The timetable outlined in Table 4.1 breaks the process down into stages, based on a one-week time frame. You can adjust the time frame according to your own deadlines.

Format Your Report Correctly

Although content is important, the appearance of your paper is what makes the first impression on the reader. If the pages are out of order and the ink is faded, subconsciously or not, the reader is going to associate a sloppy paper with sloppy science. You cannot afford that kind of reputation. In order for your work to be taken seriously, your paper has to have a professional appearance.

Scientific journals specify the format in their "Instructions to Authors" section. If your instructor has not given you specific instructions, the layout specified in Table 4.2 will give your paper a professional look.

TABLE 4.1 *Continued*

TIME FRAME	ACTIVITY	RATIONALE
Day 6	Peer reviewer reviews laboratory report.	The peer reviewer should review the paper according to two sets of criteria. One is the conventions of scientific writing as described in "Scientific Paper Format," and the other is the set of questions on the "Peer Review Guide" (see Appendix 3).
	Hold writing conference during which the reviewer returns the first draft to the writer.	An informal discussion between the writer and the reviewer is useful to give the writer an opportunity to explain what he/she intended to accomplish, and for the reviewer to provide feedback.
Days 6–7	Revise laboratory report.	Based on your discussion with your reviewer, revise as necessary. Remember that you do not have to accept all of the reviewer's suggestions.
Day 8	Hand in both first draft and revised draft to instructor (plus draft self-assessment and peer reviewer's comments).	Your instructor wants to know what you've learned (we never stop learning either!).

Consult the sample "good" student laboratory report in Chapter 6 for an overview of the style and layout. For details on how to format documents in Microsoft® Word, see Appendix 1 "Word Processing Basics."

Computer Savvy

Know your PC and your word processing software. Most of the tasks you will encounter in writing your laboratory report are described in Appendix 1 "Word Processing Basics" and Appendix 2 "Making XY Graphs in Excel." If there is a task that is not covered in these appendices, write it down and ask an expert later. If you run into a major problem that prevents you from using your PC, you should have a backup plan in place (familiarity with another PC).

Always back up your files on floppy disks or zip disks. Save your file frequently while writing your paper. Set up your computer to auto-

TABLE 4.2 Instructions to authors of laboratory reports

FEATURE	LAYOUT
Paper	8½" x 11" (or DIN A4) white bond, one side only
Margins	1½" left and right; 1" top and bottom
Font size	12 pt (points to the inch)
Typeface	Times Roman or another *serif* font. A serif is a small stroke that embellishes the character at the top and bottom. The serifs create a strong horizontal emphasis, which helps the eye scan lines of text more easily.
Symbols	Use word processing software. Do not write symbols in by hand.
Pagination	Arabic number, top right on each page except the first
Justification	Align left/ragged right or Full/even edges
Spacing	Double, except title, list of authors, and figure and table titles (which should be single-spaced)
New paragraph	Indent 0.5"
Title page (optional)	Title, authors (your name first, lab partner second), class, and date
Headings	Align headings for Abstract, Introduction, Materials and Methods, Results, Discussion, and References on left margin or center them. Use consistent format for capitalization. Do not start each section on a new page unless it works out that way coincidentally.
Subheadings	Use sparingly and maintain consistent format.
Tables and figures	Incorporate into text as close as possible after the point where they are first mentioned. Use descriptive titles, sequential numbering, proper position above or below graphic. May be attached on separate pages at end of document, but must still have proper caption.
Sketches	Hand-drawn in pencil or ink. Other specifications as in "Tables and figures," above.
References	Citation-Sequence System: Make a numbered list in order of citation. Name-Year System: List references in alphabetical order by the first author's last name. Use a hanging indent (all lines but the first indented) to separate individual references. Both systems: Use accepted punctuation and format.
Assembly	Place pages in order, staple top left.

matically save data every 10 minutes, so that in the event of a power failure you will have lost only 10 minutes' worth of work. These tasks are described in Appendix 1.

Install antivirus software on your computer, and always check floppy disks for viruses before you use them. Beware of files attached to e-mail messages. Do not open attachments unless you are sure they come from a reliable source.

Store floppy disks and zip disks in their protective boxes. Keep them away from magnetic devices (TV, speakers, etc.) and excess humidity, heat, and cold.

If you must eat and drink near a computer, keep beverages and crumbs away from the hard drive and keyboard.

Getting Started

Set aside 1 hour to begin writing the laboratory report as soon as possible after doing the laboratory exercise. Restricting the time to 1 hour forces you to make effective use of limited time. It also takes advantage of the attention span to which you are accustomed in lecture.

If your paper is progressing well and you are having fun, extend the 1-hour period until you perceive a loss of concentration. Promise yourself a reward after a certain amount of progress. It is not necessary to deprive yourself of pleasure as long as you work efficiently.

Reread the Laboratory Exercise

You cannot begin to write a paper without a sense of purpose. What were the objectives of your experiment? What questions are you supposed to answer? Take notes on the laboratory exercise to prevent problems with plagiarism when you write your laboratory report.

Audience and Tone

Scientific papers are written for scientists. Similarly, laboratory reports are intended to describe procedures and findings to fellow students. Of course, your instructor is going to read and evaluate your laboratory report, but your instructor is not the audience for whom you are writing. You are *writing for your peers*—students just like yourself.

Your audience has a knowledge base similar to your own. Thus, when you introduce your topic and describe and interpret your results, you should assume that your audience will know some scientific vocabulary, but that you should clarify or define less familiar terms. When deciding on how much background information to include, assume that your audience knows what you learned in class.

Keep the tone of your laboratory report factual and objective. Do not use jargon (terms known only to experts) or copy information verbatim from journal articles or your textbook. Remember that your objective is to write for your peers in a style that they can understand clearly.

Start with the Materials and Methods Section

The order in which you write the different sections is not the order in which they appear in the finished laboratory report. The rationale for this plan will become obvious as you read on.

The Materials and Methods section requires the least amount of thought, because you are primarily restating the procedure in your own words. Laboratory exercises are typically written in present tense, with instructions in the form of commands. When you write your laboratory report, however, summarize what you did in **full sentences and well-developed paragraphs**. Use **past tense,** because you completed the experiment some time ago.

Write the procedure in the **passive voice** if your presence was not crucial to the procedure. For example:

I peeled the potatoes and put them in the blender.

should be written:

The potatoes were peeled and homogenized.

because who did it is irrelevant to the procedure. Passive voice shifts the emphasis from "I" to "the potatoes," where the emphasis really belongs. See "Tense" in Chapter 5 for appropriate use of active and passive voice.

When you first begin to write scientific papers, it may be hard to decide how much detail to include in the Materials and Methods section. On the one hand, too many details bore the reader; on the other hand, you must include enough information to allow the reader to reproduce the experiment. Assume that the reader has the same knowledge base as you have. Because you have been instructed in certain basic laboratory techniques, it is not necessary to describe conventional labware, calculations for making solutions, operating instructions for instruments, and similar details.

Here is an example of **too much detail**:

The enzyme catalase needed to be extracted from the potato tuber. To isolate the catalase, the potato piece was peeled, cut into cubes, and weighed out on a scale until there was about

50 g of potato tissue. Next, the tissue was placed in 50 ml of cold, distilled water with a small amount of crushed ice and placed into a cold blender. These were all blended together for about 30 seconds on high speed. Immediately the mixture was placed into a graduated cylinder and placed in an ice bath until the mixture could be filtered. When a filter was available, filter paper was moistened with water and placed into the Buchner funnel. The funnel was attached to a clean filter flask that was attached to an aspirator, which created a vacuum that pulled the water and catalase from the potato mixture. This filtrate, which was collected in the filter flask, was then placed in a 100-ml graduated cylinder and filled to the 100-ml mark with cold, distilled water. This served as the original enzyme concentration for which the catalase concentrations were prepared.

The preceding example could be shortened, without losing any essential information, to:

In order to extract catalase from potatoes, about 50 g of peeled potato was homogenized with 50 ml of cold, distilled water. The mixture was filtered, and the filtrate was brought up to 100 ml with cold, distilled water. The concentration of this material was 100 units of enzyme/ml.

On the other hand, here is an example of **not enough detail**:

In this lab we used a potato, a blender, distilled water, crushed ice, and ample amounts of filter paper. The extract we started with had 100 units of enzyme/ml.

This procedure does not give the reader enough information to reproduce the experiment, because essential details like *how much* potato, *how much* water, and the definition of 100 units of enzyme/ml are left out.

The separation of the heading "Materials and Methods" into two parts is understandably confusing for beginning students. The implication is that the materials should be listed separately from the methods. In fact, **materials should not be listed separately** unless the bacteria strain, vector (plasmid), growth media, or chemicals were obtained from a special or noncommercial source. It will be obvious to the reader what materials are required on reading the methods.

If you are paraphrasing a published laboratory exercise, it is necessary to cite the source (see "References," p. 41). Unpublished laboratory exercises are not usually cited; ask your instructor to be sure.

Do the Results Section Next

Reread the questions in the laboratory exercise to determine what your instructor expects you to learn from the data. You may also find specific instructions on how to organize the data (tables and graphs). Based on these expectations, decide how best to reduce (organize) the raw data. Ask yourself the following questions:

- Can I state the results in one sentence? If so, then neither a graph nor a table is needed.
- Are the numbers themselves more important than the trend shown by the numbers? If so, then use a table.
- Is the trend more important than the numbers themselves? If so, use a graph.

The purpose of the Results section is to *summarize* the key findings of your experiment. It is your job as the author to analyze the raw data and to display them in a manner that provides strong support for your arguments. Reducing the raw data is really an exercise in organization. Once you have a clear overview of the findings, you can start to think about the implications.

Organizing the raw data can be difficult when the results obtained are variable or unexpected. Statistical analysis may help you decide whether the variability is inherent in the population sampled, due to the imprecision in making measurements, or due to human error. Furthermore, there may be *no* difference between the control and the experimental treatments, although you expected to find one. How do you organize the data when there are so many uncertainties?

Although there is no single right answer to this question, the answer must be based on honesty. Present the data as honestly and objectively as you can. Explain what you did and why you did it. The reader may disagree with your reasoning, but at least you were honest about your intentions.

Tables

Tables are used to display large quantities of numbers and other information that would be tedious to read in prose. Arrange the categories vertically, rather than horizontally, as this arrangement is easier for the reader to follow (Table 4.3). Notice that each column heading is followed by the units. This arrangement saves you the trouble of writing the units after each number entry in the table.

Give each table a caption that includes a number and a title. Single-space the caption, and either center it or align it on the left margin *above*

TABLE 4.3 Temperature requirements and temperature tolerance of *Microcystis* and *Synechococcus* isolates

ORGANISM	PREVAILING LIGHT INTENSITY (µE/M2/S)	LOWER TEMPERATURE LIMIT (°C)	UPPER TEMPERATURE LIMIT (°C)	OPTIMAL GROWTH TEMPERATURE RANGE (°C)	THERMAL GROWTH OPTIMUM (°C)
M. UV-001	20	12.0	36.0	24.0–34.0	29.0
M. UV-003	20	10.5	36.6	26.0–33.0	29.5
M. UV-006	20	13.5	40.0	26.0–34.5	30.3
M. UV-007	20 / 33	13.2 / 12.0	36.4 / 35.0	26.0–35.0 / 25.5–32.0	30.5 / 28.8
S. UV-005	25 / 33	11.0 / 10.3	not det. / 44.3	not det. / 26.0–43.0	not det. / 34.3

the table. Use Arabic numbers, and number the tables consecutively in the order they are discussed in the text. Notice that in this book, the table and figure numbers are preceded by the chapter number. This system helps orient the reader in long manuscripts, but is not necessary in short papers like a laboratory report.

Titles consist of a precise noun phrase, not a complete sentence. The reader should be able to understand the essence of the table without having to refer to the body (the text) of the Results section.

In your laboratory report, it is not necessary to include a table when you already have a graph that shows the same data. Make *either* a table *or* a graph—not both—to present a given data set.

There are a number of ways to construct tables in Microsoft® Word (see Appendix 1).

Figures

A figure is any graphic illustration that is not a table. Thus, line graphs, bar graphs, pie charts, drawings, gel photos, X-ray images, and microscope images are all called *figures* in scientific papers.

Line graphs are used to display a trend or an important relationship between one or more parameters. By convention, the independent variable (the one the investigator manipulates) is plotted on the x-axis, and the dependent variable (the one that changes in response to the independent variable) is plotted on the y-axis. The data then show the effect of x on y, or y as a function of x.

Figures are always numbered and titled *beneath* the graphic (Figure 4.1). The captions may be centered or placed flush on the left margin of the report. Captions are always single spaced. Arabic numbers are used, and the figures are numbered consecutively in the order they are discussed in the text. There are a variety of acceptable uses of periods, indentation, and capitalization in the figure caption. The most important thing is to be consistent. Figure 4.1 is a sample figure from a student lab report. Its caption follows *The EMBO Journal* style.

Figure titles should consist of a precise noun phrase, not a complete sentence. The reader should be able to understand the title without referring to the text in the Results section. Titles that merely state the *y*-axis label versus the *x*-axis label are *not* acceptable. If there is more than one data set (line) on the figure, a legend must be given. The legend may be included in the figure, in which case it is boxed and contained within the figure frame. Alternatively, the legend may be clarified as part of the figure caption. Either format is acceptable in scientific papers, as long as you use it consistently.

Figures in your laboratory report should be prepared according to the guidelines specified by the Council of Science Editors (formerly called the Council of Biology Editors). Although you may wish to plot a rough draft of your graphs by hand, you should learn how to use computer plotting software to make graphs for your scientific papers.

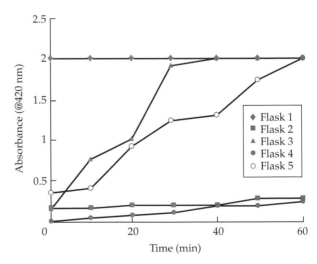

Figure 4-1 Beta-galactosidase production in *E. coli* grown under five different conditions. Flask 1 contains the lacI⁻ strain in glycerol. Flasks 2–5 contain the lacI⁺ strain. Flasks 2 and 3 contain glycerol medium; flasks 4 and 5 contain glucose. IPTG was added to flasks 3 and 5 at $t = 0$ min.

Microsoft® Excel is a good plotting program for novices (see Appendix 2) because it is readily available and fairly easy to use. The time you invest now in learning to plot data on the computer will be invaluable in your upper-level courses and later in your career.

Equations

Equations are neither tables nor figures. They should be set off from the rest of the text on a separate line. If you have several equations and need to refer to them unambiguously in the body of the Results (or other) section, number each equation sequentially and place the number in parentheses on the right margin. For example,

$$\text{Absorbance} = -\log T \tag{1}$$

If you are presenting a sequence of calculations, align the = symbol in each line, as in the following example.

Protein concentration of the unknown sample was determined using the equation of the Biuret standard curve. The measured absorbance value was substituted for y, and the equation was solved for x (the protein concentration):

$$y = 0.0417x$$
$$0.225 = 0.0417x$$
$$5.40 = x$$

Thus, the protein concentration of the sample was 5.40 mg/mL.

Body (Text) of the Results Section

The Results section has two components:

- Graphic illustrations (tables and/or figures)
- A text, in which you describe the findings presented in the graphic illustrations. The findings are described in an objective manner, without explaining why or discussing possible implications.

You started working on the Results section by organizing your raw data into tables and/or figures. The next step is to decide in which order to present the tables and figures, and then to describe each one in turn to the reader. In the text of the Results section, you state each important finding and refer the reader to the table or figure that supports this finding. An example of a **good paragraph** in the Results section is as follows:

Beta-galactosidase production varied depending on the culture conditions (Figure 1). In flask 1, which contained the mutant strain of *E. coli,* beta-galactosidase was always produced. In flasks 2 and 3, which both contained normal strains of *E. coli* as well as glycerol in the medium, beta-galactosidase production was determined by the presence or absence of IPTG in the medium.

Notice that the important result shown by Figure 1 is stated in the opening sentence of the paragraph. The figure number is enclosed by parentheses following the opening sentence, so that the reader can see the data for himself or herself. The subsequent sentences in the paragraph then provide specific details on the data displayed in the figure.

An example of a vague, uninformative, and **unsuitable sentence** is as follows:

Refer to the figures for the results.

This sentence is unsuitable because there is no guidance from the author concerning *what* findings were important, and it is unclear *how* the figures support the findings.

As you describe the important findings of your investigation, think about what you already know about the subject. Are the results expected? Do they agree with the findings of other investigators?

- If **yes**, then you can jot down your ideas to use in the Discussion section.

- If **no**, try to develop possible explanations for the results. Some reasons to consider are:
  - Human error (failure to follow the procedure, failure to use the equipment properly, failure to prepare solutions correctly, variability when multiple lab partners measure the same thing, and simple arithmetic errors). If you suspect that human error may have influenced your results, it is important to acknowledge its contribution.
  - Numerical values were entered incorrectly in the computer plotting program.
  - Sample size was too small.
  - Variability was too great to draw any conclusions.

If you can rule out these possibilities, discuss your results with your lab partner, teaching assistant, or instructor. If there is an obvious error, an "outsider" may be able to spot it immediately. Furthermore, informal discussions may help you clarify what you know and what you do not know about the topic.

Make Connections

Now that the "meat" of your report is done, it's time to describe how your work fits into the existing body of knowledge. These connections are made in the Discussion and Introduction sections.

Write the Discussion

The Discussion section gives you the opportunity to **interpret your results and explain why they are important**. For a strong Discussion section:

- Summarize the results in a way that provides evidence for your conclusions.
- State how your results relate to existing knowledge (cite literature sources).
- Point out any inconsistencies in your data. This is preferable to concealing an anomalous result.
- Discuss sources of error.
- Describe future extensions of the current work.

Write the Introduction

The Introduction concisely states what motivated the study, how it fits into the existing body of knowledge, and the objectives of the work. The introduction consists of two primary parts: (1) background information from the literature and (2) objectives of the current work.

After having written drafts of the Materials and Methods, Results, and Discussion sections, you should be intimately familiar with the procedure, the data, and what the data mean. Now you are in a position to put your investigation into perspective. What was already known about the topic? Were there any inconsistencies or unanswered questions? Why did you carry out this investigation?

If you designed your own experiment based on the scientific method, you probably answered these questions already. If this investigation was a laboratory exercise prepared by your instructor, reread the exercise to see if some of these questions have already been answered. The idea is not to copy the laboratory exercise introduction verbatim, but to make sure you understand the objectives.

Use keywords from the objectives to locate background information in the literature (see Chapter 2). Evaluate the titles in the "hits" critically to make sure the sources are relevant to your investigation.

The opening sentence of the Introduction section is usually a general observation or result familiar to the audience. Subsequent sentences narrow down the topic to the specific focus of the current investigation. Subsequent paragraphs then provide background information from the literature and describe unanswered questions or inconsistencies. The objectives of the current work are usually stated in the last paragraph of the Introduction section.

Effective Advertising

The whole point of writing your paper is to communicate your work to your peers. The Abstract and the Title are the primary tools your audience will use to decide whether or not they are interested in your work.

Write the Abstract

The abstract is a **summary of the entire paper** in 250 words or less. It contains:

- An introduction (scope and purpose)
- A short description of the methods
- The results
- Your conclusions

There are no literature citations or references to figures in the Abstract.

After the title, the Abstract is the most important part of the scientific paper used by the audience to determine initial interest in the author's work. Abstracts are indexed in databases that catalogue the literature in the biological sciences. If an abstract suggests that the author's work may be relevant to your own work, you will probably want to read the whole article. On the other hand, if an abstract is vague or essential information is missing, you will probably decide that the paper is not worth reading. When you write the Abstract for your own laboratory report, put yourself in the position of the reader. If you want the reader to be interested in your work, write an effective Abstract.

Writing the Abstract is difficult because you have to condense your entire paper into 250 words or less. One strategy for doing this is to list the key points of each section, as though you were taking notes on your own paper. Then write the key points in full sentences. Revise the draft for clarity and conciseness using strategies such as using active voice, combining choppy sentences with connecting words, rewording run-on sentences, and eliminating redundancy. With each revision, look for ways to shorten the text so that the resulting Abstract is a concise and accurate summary of your work.

The ability to write abstracts is important to a scientist's career. Should you someday wish to present your research at an academic society meeting, such as the Society for Neuroscience, the American Association for the Advancement of Science, or the National Association of Biology Teachers (to name just a few), you will be asked to submit an abstract of your presentation to the committee in charge of the meeting program. Your chances of being among the select field of presenters at these meetings are much better if you have learned to write a clear and intelligent abstract.

Write the Title

The title is a short, informative description of the **essence of the paper**. You may choose a working title when you begin to write your paper, but revise the title after subsequent drafts. Remember that readers use the title to determine initial interest in the paper, so descriptive accuracy is the most essential element of your title. Brevity is nice if it can be achieved. Some journals (especially the British ones) are also fond of puns and humor in their titles, but this kind of thing may be better left for later in your career.

References

References are typically cited in the Introduction and Discussion sections of a scientific paper, and the procedures given in Materials and Methods are often modifications of those in previous work. Citation and reference format in the sciences differs from that in the liberal arts in three important ways:

- It is not customary to use quotation marks in scientific papers; instead, paraphrase.
- In the citation in the text, it is not customary to give the page number(s) of the source; instead, use one of the two citation formats explained in the following section.
- In the references at the end of the report, give the page numbers for the entire paper (inclusive page numbers), not just the page(s) from which you got information.

The References section is also called "Literature Cited," because it includes only the published sources you cited in your paper. It is not a "bibliography," which is a list of all the works you consulted to learn more about the topic.

Citation Format

The Council of Biology Editors manual, *Scientific Style and Format* (6th ed.), recommends the following two formats. The one you actually use depends on your instructor's preference or on the format specified by the particular scientific journal in which you aspire to publish.

1. **Citation-Sequence System.** In the text, the citation is given by a number in square brackets or parentheses following the cited sentence (some journals require this number to be a superscripted endnote, but never a footnote). On the page of references that follows the Discussion section, the citations are listed in **numerical order** and include the full reference.

   Example of citation:

   ...was confirmed in a previous study (1).

   Example of reference:

   1. Ishikawa H, Evans M. Specialized zones of development in roots. Plant Physiology 1995; 109: 725–727.

2. **Name-Year System.** In the text, the citation is given in the form of author(s) and year in parentheses following the cited sentence. On the page of references that follows the Discussion section, the references are listed in **alphabetical order** according to the first author's last name.

   Examples of citations:

   - If the paper has one or two authors, cite the last name(s):

     ...was confirmed in a previous study (Ishikawa and Evans, 1995).

   - If the paper has three or more authors, cite only the last name of the first author and follow this with "and others" or "et al." The Latin abbreviation et al. (for *et alia*, "and others") is used in most journal articles, whereas the CBE Manual prefers the English equivalent (CBE Manual, 1994).

     ...was confirmed in a previous study (Curtright et al. 1996).

     or

     ...was confirmed in a previous study (Curtright and others, 1996).

■ If you cite more than one paper published by the same
   author in the same year, add a letter after the year:

   … was described in recent work by Dawson (1999a, 1999b).

The name-year system has the advantage that people working in the
field will know the literature and, on seeing the authors' names, will
understand the reference without having to check the reference list. With
the citation-sequence system, the reader must turn to the reference list at
the end of the paper to gain the same information. The important thing
is to choose a citation format you are comfortable with and adhere to it
consistently throughout your paper.

Personal Communication

If you had a discussion with your professor or colleague, and you
obtained certain information from him or her that you could not find in
a published source, it is appropriate to recognize this person's contribution as follows:

   …may be explained by possible contamination from a virus or
   bacterium (Pizzorno, personal communication).

It is **not necessary** to include personal communications in the References
section.

Reference Formats

**Only references that have been cited in the text may be included in
the References (also called Literature Cited) section.** Some examples
of how to write the full reference for a journal article, an article written
in a book, a book, an unpublished laboratory exercise, and an article on
the Internet are given on the following pages. The last name is written
first, followed by the initials. When there are 10 or fewer authors, list all
authors' names. When there are more than 10 authors, list the first 10
and then write "et al." or "and others" after the tenth name.

The only difference between the Citation-Sequence System (C-S) and
the Name-Year System (N-Y) is the sequence of information in the reference. In the Name-Year System, the publication year follows the
author's name; in the Citation-Sequence System, the publication year
follows the journal name.

The general format, sequence of information, and punctuation advo-
cated by the CBE Manual are shown on the following tabbed page. In the
N-Y system, a hanging indent (as shown in these sample formats) is
used in all lines after the first to separate the individual references.

Journal article

C-S Number of the citation. All authors' names separated by commas. Article title. Journal title year month; volume number(issue number): inclusive pages.

N-Y All authors' names separated by commas. Year of publication. Article title volume number(issue number): inclusive pages.

EXAMPLE: journal article, two authors

C-S 1. Ishikawa H, Evans M. Specialized zones of development in roots. Plant Physiology 1995; 109: 725–727.

N-Y Ishikawa H, Evans M. 1995. Specialized zones of development in roots. Plant Physiology 109: 725–727.

EXAMPLE: journal article, three or more authors

C-S 1. Curtright R, Rynearson JA, Markwell J. Anthocyanins. J. Chem. Educ. 1996; 73(4): 306–310.

N-Y Curtright R, Rynearson JA, Markwell J. 1996. Anthocyanins. J. Chem. Educ. 73(4): 306–310.

Note: In some journals, the year of publication is placed in parentheses. Sometimes a comma is used instead of a colon after the volume and issue number.

Article in book

C-S Number of the citation. All authors' names separated by commas. Article title. In: Editors' names, editors. Book title. Edition. Publisher's location: Publisher's name; Year of publication. pp inclusive pages.

N-Y All authors' names separated by commas. Year of publication. Article title. In: Editors' names, editors. Book title. Edition. Publisher's location: Publisher's name. pp inclusive pages.

EXAMPLE

C-S 1. Dennison DS. Phototropism. In: Haupt W, Feinleib ME, editors. Encyclopedia of Plant Physiology, New Series, Vol 7, Physiology of Movements. Berlin: Springer Verlag; 1979. pp 506–566.

N-Y Dennison DS. Phototropism. In: Haupt W, Feinleib ME, editors. Encyclopedia of Plant Physiology, New Series, Vol 7, Physiology of Movements. Berlin: Springer Verlag. pp 506–566.

Note: In some journals, the year of publication is placed in parentheses. Sometimes a comma is used instead of a colon between the city and publisher's name.

Book

C-S   Number of the citation. All authors' names [or editor(s)]. Title of book. Publisher's location: Name of publisher; Year. Number of pages.

N-Y   All authors' names [or editor(s)]. Year of publication. Title of book. Publisher's location: Name of publisher. Number of pages.

EXAMPLE

C-S   1. Treshow M. Environment and plant response. New York: McGraw-Hill. 1970. 250 p.

N-Y   Treshow M. 1970. Environment and plant response. New York: McGraw-Hill. 250 p.

Note: In some journals, the year of publication is placed in parentheses. Sometimes a comma is used instead of a colon between the city and publisher's name. Sometimes the order of city and publisher's name is reversed.

Unpublished Laboratory Exercise

Unpublished material is usually not included in the References section. If your instructor asks you to cite the laboratory exercise in your laboratory report, however, the format could look like this:

C-S   Author ([Anonymous] if unknown). Title of lab exercise. Course number, Department, University. Year.

N-Y   Author ([Anonymous] if unknown). Year. Title of lab exercise. Course number, Department, University.

EXAMPLE

C-S   1. [Anonymous]. Isolation of Mitochondria by Differential Centrifugation. BIOL202, Biology Department, Bucknell University. 2000.

N-Y   [Anonymous]. 2000. Isolation of Mitochondria by Differential Centrifugation. BIOL202, Biology Department, Bucknell University.

Reference Formats

Article on the Internet

C-S   1. Jacobson JW, Mulick JA, Schwartz AA. A history of facilitated communication: Science, pseudoscience, and antiscience: Science working group on facilitated communication. American Psychologist 1995; 50: 750–765. Retrieved January 25, 1996 from the World Wide Web: <http://www.apa.org/journals/jacobson.html>

N-Y   Jacobson JW, Mulick JA, Schwartz AA. 1995. A history of facilitated communication: Science, pseudoscience, and anti science: Science working group on facilitated communication. American Psychologist 50: 750–765. Retrieved January 25, 1996 from the World Wide Web: http://www.apa.org/journals/jacobson.html

Sources from the Internet and the World Wide Web

Ease of access to the latest information makes the World Wide Web an irresistible source for nearly everything. Scientific communications are no exception. Take heed, however; information retrieved from the Internet and the World Wide Web may be unreliable. The authors of the communication, not journal editors or copy editors, are responsible for the accuracy of the contents. Because there is no review from an outside source, mistakes can occur. For this reason, it is preferable to cite only reliable published books or articles in your laboratory report.

In some instances, sources from the Web may be used to support published findings. Use one of the two citation formats described previously. If the author is not known, simply abbreviate the title of the Web page and include the date. For example:

... is recommended by the Counselling Services at the University of Victoria (How to Read University Texts, 2001).

When URLs are used in text, they are enclosed in angle brackets (< >) to distinguish them from the rest of the text. Every character is significant, as are spaces and capitalization. Very long URLs can be broken before a punctuation mark (tilde ~, hyphen -, underscore _, period ., forward slash /, backslash \, or pipe |). The punctuation mark is then moved to the next line, as in the following example:

http://www.coun.uvic.ca/learn/program/hndouts/Readtxt.html

In the References section, begin with the same information that would be provided for a printed source (or as much of that information as available). The Web information is then placed at the end of the reference. It is *absolutely essential* to use "Retrieved from" and the date, because documents on the Web may change in content, move, or be removed from a site altogether. The last entry on the Reference Format pages shows how to cite an Internet article.

For more information on using Internet sources, see Harnack and Kleppinger (2001).

Chapter 5

REVISION

Revision—reading your paper and making corrections and improvements—is an important task that usually does not get nearly the attention it deserves. Too many students write the first draft of their laboratory report the night before it is due and hand in the hard copy, still warm from the printer, without even having proofread it.

The truth is, most writers cannot produce a clear, concise, and error-free product on the first try. It may take several revisions before the writer is satisfied that he or she has conveyed, with clarity and logic, the motivation for writing the paper, the important findings, and the conclusions. For this reason, **prepare the first draft as early as possible** so that you will have the time to think about it, get help if necessary, get feedback from your peer reviewer, and make final revisions.

Getting Ready to Revise

Take a Break

The first step in revision is *not* to do it immediately after you have written the first draft. You need to distance yourself from the paper to gain the objectivity needed to read the paper critically. So take a break, and get a good night's sleep.

Look at the Big Picture

In the early stages of the revision process, content is more important than format. Make sure you have answered all the questions in the laboratory exercise (if applicable), and use the "Revision Checklist" to make sure each section of your laboratory report contains the proper content. Don't worry about whether or not a table caption is in the right position until you are sure you even need to include the table.

Ask yourself the following questions:

- What are the goals and objectives of this paper? Did I make these objectives clear to the reader?
- What questions and concerns do I have with this draft? What are some ways I can begin to address these concerns?
- What parts need feedback (not sure if my meaning is clear, not sure if I understood this concept correctly, and other issues)?
- What do I like about this paper? What are its strengths and what has gone well?

Answer these questions on the "Draft Self-Assessment Form" (Appendix 3). This form alerts your peer reviewer (peer review is the next step in the revision process) to your goals and concerns with your paper.

Get Feedback

When we are engrossed in our work, we may not recognize that what is obvious to us is not obvious to an "outsider." Typical problem areas are flow and organization of the paper and typographical or grammatical errors.

That is where feedback from someone who is familiar with the subject matter comes in handy. Ask your lab partner or another classmate to review your paper. Return the favor by reviewing his or hers. The questions your reviewer will focus on are as follows:

- Do I know what the writer is trying to accomplish with this paper? Is the purpose clear?
- What questions or concerns do I have about this paper? Are there sections that were difficult to follow? Are the organization, content, flow, and level appropriate for the intended audience?
- What suggestions can I offer the writer to help him/her clarify the intended meaning?
- What do I like about the paper? What are its strengths?

By answering these questions on the "Peer Review Form" (Appendix 3), the reviewer provides the writer with a tangible impression of the overall effectiveness of the paper.

Tips for being a good peer reviewer.   There are two issues with which you may struggle when you are asked to review your classmate's paper: (1) I'm not confident that I know the "right" answer or know enough about the writing process to give good suggestions, and (2) I don't want to hurt the writer's feelings. These are valid concerns, and resolving them will require, first, a willingness to learn as much about writing scientific papers as possible, and second, the attitude that if something is unclear to you, it may also be unclear to other readers. With each paper you review, you will gain more confidence in your ability to give constructive feedback. In the meantime, however, a good rule of thumb is to give the kinds of suggestions and consideration that you would like to receive on your own paper.

There are two schools of thought concerning how the reviewer should provide feedback on the paper. One is that no marks should be made on the paper itself; instead, any comments should be written on a separate sheet of paper. The second school of thought is just the opposite. Although the first method respects the writer's ownership of the paper, it is not very practical. First of all, it is difficult to describe exactly where on the paper the revision should be made. Secondly, vague suggestions like "improve flow of introduction" are not nearly as helpful to your classmate as "what do you mean by..." and then circle or underline the confusing phrase. Think of the peer review process as a team sport: you are not challenging your teammate's right to be on the team. You are working together to get the best possible result.

If it is not possible for you and your classmate to exchange lab reports in person, you might consider sending your paper by e-mail as an attached file. To see the changes made by your peer reviewer, use the Track Changes feature in Word (See "Tracking Changes made by Peer Reviewers" in Appendix 1).

Here are some concrete tips for being a good peer reviewer:

- Read the writer's draft self-assessment prior to reading the paper.
- Use the Revision Checklist for content.
- Use a pencil and standard proofreader's marks to mark awkward sentences, spelling and punctuation mistakes, and formatting errors. Do not feel you have to rewrite individual sentences—that is the writer's job.
- Ask questions. Let the writer know where you can't follow his/her thinking, where you need more examples, where you expect more detailed analysis, and so on.
- Do not be embarrassed about making lots of comments; the author does not have to accept your suggestions. On the other hand, if you say only good things about the paper, how will the writer know whether the paper is accomplishing the desired objectives?

You can fine-tune your proofreading skills on any text. You may recognize some of your own problems in other people's writing, and with persistence and practice, you will find creative solutions to correct these problems. Keep a log of the problems that recur in your writing and review them from time to time. Repetition builds awareness, which will help you achieve greater clarity in your writing.

Have an informal discussion with your peer reviewer. Sometimes the comments made by the peer reviewer on the paper are self-explanatory. Other times, however, the peer reviewer cannot respond to certain parts of the paper, because more information is required. Under these circumstances, an informal discussion between the writer and the reviewer is helpful. There are two important rules for this discussion:

- First, the writer talks and the reviewer listens. The objective is to help the writer express exactly what he/she wants to say in the paper.
- Second, the reviewer talks, in nonjudgmental terms, about which parts of the paper were readily understandable and which parts were confusing. The reviewer does not have to be an experienced writer to do this—no two people have exactly the same life experiences, and there is always something positive you can learn from looking at your writing from someone else's perspective.

Revise Again

The comments provided by your peer reviewer, both on your paper and during the subsequent informal discussion, should give you some ideas on how your paper can be improved. The areas that require attention will most likely fall into one of the following categories:

- Conventions in biology
- Numbers
- Standard abbreviations
- Punctuation
- Clarity
- Grammar
- Word usage
- Spelling
- Global revision

In Appendix 4, examples of poorly written sentences are organized according to these categories. If you notice that you consistently make a certain kind of mistake when you write, practice rewriting some of the sentences in that particular category.

Conventions in Biology

Audience. Scientific papers are written by scientists for fellow scientists. When you write your laboratory reports, write for your fellow students. Even though your instructor will be the one who reads and evaluates your reports, think of him or her as a mentor, not the audience.

The audience has a knowledge base similar to the author of the paper. Thus, when you introduce your topic and describe and interpret your results, you should assume that your audience will know some scientific vocabulary, but that you should clarify or define less-familiar terms. When deciding on how much background information to include, assume that your audience knows what you learned in class.

Do not use jargon (terms known only to experts) or copy sections of journal articles because you think that your own words are inferior to those of experts in the field. The objective of your writing is not to impress your instructor with empty words, but to enable your peers to understand what you have accomplished.

Introductions. In some fields, the author attempts to attract a large audience for his or her paper by asking an intriguing question or presenting a dilemma to capture the reader's interest. In scientific papers, however, the author assumes that the audience is already interested in the subject. Readers of scientific papers typically determine initial interest in a paper from the title and the abstract. If these sections seem promising, the reader will then peruse the Introduction to determine further interest. There is no need for the author to pique the reader's interest with longwinded introductions, because the interest is already there.

When you write the Introduction to your laboratory report, just jump right in with your topic. Do not begin with unnecessary introductions like:

> Enzymes are interesting biological molecules.

or

> In many areas of biology, it is important to gain a higher understanding of enzyme activity.

Any reader who was sufficiently interested in your title and abstract to read the Introduction section knows this. Get to the point. Describe what area of enzyme activity you studied, why, and what you expected to learn from your work.

Initially it is difficult to write in (and read) the terse, get-to-the-point style that characterizes scientific papers. With practice, however, you may come to appreciate this style, because in a well-written paper, not a word is wasted. The benefit to you as a reader is that you extract a maximum amount of information from a minimum amount of text.

Present tense or past tense? In scientific papers, present tense is used mainly in the following situations:

- To make generally accepted statements (for example, "Photosynthesis *is* the process whereby green plants produce sugars.")
- When referring directly to a table or figure in your paper (for example, "Figure 1 *is* a schematic diagram of the apparatus.")
- When you state the findings of published authors (for example, "Catalase HPII from *E. coli is* highly resistant to denaturation (Switala and others, 1999).")

Past tense is used mainly in the following situations:

- To report your own work, especially in the Abstract, Materials and Methods, and Results sections, because it remains to be seen if it is accepted as fact (for example, "At temperatures above 37°C, catalase activity *decreased* (Fig. 1).")
- To cite another author's findings directly (for example, "Miller and others (1998) *found* that…")

Active and passive voice. In **active voice**, the subject does the action. In **passive voice**, the subject receives the action. Consider the following examples:

PASSIVE: Catalase was extracted from a potato.
           [Emphasis on *catalase*]

ACTIVE: I extracted catalase from a potato.
          [Emphasis on *I*]

Notice the difference in emphasis. Is it really important to the success of the procedure that *you* did it, or does the emphasis belong on the materials? Use of personal pronouns in scientific writing is often a matter of personal preference or of following the trend in a particular specialty area. As a general rule, however, write in passive voice whenever your presence is not essential to the success of the action.

Another use of active and passive voice is to express your opinion, as in the following example:

ACTIVE: I concluded from this observation that...

PASSIVE: The researcher concluded from this observation that...
It was concluded from this observation that...

In all three sentences, the subject is *I* (the author of the paper), yet only the active voice conveys this clearly. The passive voice leaves the reader wondering who is drawing the conclusion. As a general rule, use *I* when the statement is your own opinion.

Finally, when you are not the subject of the sentence, use active voice whenever possible. Consider the following observation:

ACTIVE: Brine shrimp developed faster in 30 ppt seawater than in 50 ppt.

PASSIVE: Fifty ppt seawater was less conducive to the development of brine shrimp than 30 ppt.

The sentence written in active voice is shorter and more dynamic than the sentence written in passive voice. The passive voice, in contrast, is wordy and makes the writer seem pompous. As a general rule, use active voice as much as possible whenever you are not the subject of the sentence.

Numbers

Numbers are used for quantitative measurements. In the past, numbers less than 10 were spelled out, and larger numbers were written as numerals. The modern scientific number style described in the Corrections to the CBE Manual (6th ed.) aims for a more consistent usage of numbers. The new rules are as follows:

1. Use numerals to express any *quantity*. This form increases their visibility in scientific writing, and emphasizes their importance.
   - Cardinal numbers, for example, 3 observations, 5 samples, 2 times
   - In conjunction with a unit, for example, 5 g, 0.5 mm, 37°C, 50%, 1 hr. Pay attention to spacing, capitalization, and punctuation (see Table 5.1).
   - Mathematical relationships, for example, 1:5 dilution, 1000× magnification, 10-fold

2. Spell out numbers in the following cases:
   - When the number begins a sentence, for example, *"Fifty g of potatoes was [not were] homogenized."* rather than *"50 g of potatoes was homogenized."* Alternatively, restructure the sentence

so that the number does not begin the sentence. Notice that when numbers are used in conjunction with units, the quantity is considered to be singular, not plural.

- When there are two adjacent numbers, retain the numeral that goes with the unit, and spell out the other one. An example of this is *The solution was divided into four 250-mL flasks.*
- When the number is used in a nonquantitative sense, for example, *one of the treatments, the expression approaches zero, one is required to consider…*
- When the number is an ordinal number less than 10, and when the number expresses rank rather than quantity, for example, *the second time, was first discovered.*
- When the number is a fraction used in running text, for example, *one-half of the homogenate, nearly three-quarters of the plants.* When the precise value of a fraction is required, however, use the decimal form, for example, *0.5 L* rather than *one-half liter.*

3. Use scientific notation for very large or very small numbers. For the number 5,000,000, write 5×10^{-6}, not *5 million*. For the number *0.000005*, write 5×10^{-6}.

4. For decimal numbers less than one, always mark the ones column with a zero. For example, write *0.05*, not *.05.*

Standard Abbreviations

The CBE Manual (6th ed.) defines standard abbreviations for authors and publishers in the sciences and mathematics. Some of the terms and abbreviations that you may encounter in introductory biology courses are shown in Table 5.1. Take note of spacing, case (capital or lowercase letters), and punctuation use. Except where noted, the symbols are the same for singular and plural terms (for example, 30 min *not* 30 mins).

Punctuation

The purpose of punctuation marks is to divide sentences and parts of sentences to make the meaning clear. A few of the most common punctuation marks and their uses are described in the following section. For a more comprehensive, but still concise, treatment of punctuation, see Hacker (1997), Lunsford (2001), or Lunsford and Connors (1995, 1999).

The comma.   The comma inserts a pause in the sentence in order to avoid confusion. Note the ambiguity in the following sentence:

While the sample was incubating the students prepared the solutions for the experiment.

A comma *should* be used in the following situations:

1. To connect two independent clauses that are joined by *and, but, or, nor, for, so,* or *yet.* An independent clause contains a subject and a verb, and can stand alone as a sentence.

   EXAMPLE: Feel free to call me at home,but don't call after 9 P.M.

2. After an introductory clause, to separate the clause from the main body of the sentence.

   EXAMPLE: Although she spent many hours writing her lab report, she earned a low grade because she forgot to answer the questions in the laboratory exercise.

   A comma is not needed if the clause is short.

   EXAMPLE: Suddenly the power went out

3. Between items in a series, including the last two.

   EXAMPLE: Enzyme activity is affected by factors such as substrate concentration, pH, temperature, and salt.

4. Between coordinate adjectives (if the adjectives can be connected with *and*)

   EXAMPLE: The students' original, humorous remarks made my class today particularly enjoyable. [*Original and humorous remarks* makes sense.]

   A comma is not needed if the adjectives are cumulative (if the adjectives cannot be connected with *and*).

   EXAMPLE: The three tall muscular students look like football players. [It would sound strange to say *three and tall and muscular students.*]

5. With *which,* but not *that* (see Word usage: that, which)

6. To set off conjunctive adverbs such as *however, therefore, moreover, consequently, instead, likewise, nevertheless, similarly, subsequently, accordingly,* and *finally*

TABLE 5.1 Standard abbreviations in scientific writing

TERM	SYMBOL OR ABBREVIATION	EXAMPLE
Latin words and phrases [The Council of Biology Editors Manual recommends that Latin words be replaced with English equivalents.]		[The Latin word may be replaced with the English equivalent given in brackets.]
circa (approximately)	ca.	The lake is ca. [approx.] 300 m deep.
et alii (and others)	*et al.*	Jones *et al.* [and others] (1999) found that …
et cetera (and so forth)	etc.	pH, alkalinity, etc. [and other characteristics] were measured.
exempli gratia (for example)	e.g.	Water quality characteristics (e.g., [for example,] pH, alkalinity) were measured.
id est (that is)	i.e.	The enzyme was denatured at high temperatures, i.e., the enzyme activity was zero. [Because the enzyme was denatured at high temperatures, the enzyme activity was zero.]
nota bene (take notice)	NB	NB [Important!]: Never add water to acid when making a solution.
Length		
nanometer (10^{-9} meter)	nm	*Note:* There is a space between the number and the abbreviation. There is no period after the abbreviation.
micrometer (10^{-6} meter)	μm	
millimeter (10^{-3} meter)	mm	
centimeter (10^{-2} meter)	cm	
meter	m	450 nm, 10 μm, 2.5 cm
Mass		
nanogram (10^{-9} gram)	ng	*Note:* There is a space between the number and the abbreviation. There is no period after the abbreviation.
microgram (10^{-6} gram)	μg	
milligram (10^{-3} gram)	mg	
gram	g	
kilogram (10^{3} gram)	kg	450 ng, 100 μg, 2.5 g, 10 kg

TABLE 5.1 *Continued*

TERM	SYMBOL OR ABBREVIATION	EXAMPLE
Volume		
microliter (10^{-6} liter)	μl or μL	*Note:* There is a space between the number and the abbreviation. There is no period after the abbreviation.
milliliter (10^{-3} liter)	ml or mL	
liter	l or L	
cubic centimeter (ca. 1 mL)	cm^3	450 μl or 450 μL, 0.45 ml or 0.45 mL, 2 l or 2 L
Time		
seconds	s or sec	*Note:* There is a space between the number and the abbreviation. There is no period after the abbreviation (unless the unit ends a sentence).
minutes	min	
hours	h or hr	
days	d	
		60 s or 60 sec, 60 min, 24 h or 24 hr, 1 d
Concentration		
molar (U.S. use)	M	TBS contains 0.01 M Tris-HCl, pH 7.4 and 0.15 M NaCl.
molar (SI units)	mol L^{-1}	
parts per thousand	ppt	Brine shrimp can be raised in 35 ppt seawater.
Other		
degree Celsius	°C	15°C (no space between number and symbol)
degree Fahrenheit	°F	59°F (no space between number and symbol)
diameter	diam.	pipe diam. was 10 cm
figure, figures	Fig., Figs.	As shown by Fig. 1, ...
foot-candle	fc or ft-c	500 fc or 500 ft-c
maximum	max	The max enzyme activity was found at 36°C.
minimum	min	The min temperature of hatching was 12°C.
mole	mol	
percent	%	95% (no space between number and symbol)
species (sing.)	sp.	*Tetrahymena* sp.
species (plur.)	spp.	*Tetrahymena* spp.

EXAMPLE: Instructors expect students to hand in their work on time; however, illness and personal emergencies are acceptable excuses.

7. To set off transitional expressions such as *for example, as a result, in conclusion, in other words, on the contrary,* and *on the other hand*

EXAMPLE: Chuck participates in many extracurricular activities in college. As a result, he rarely gets enough sleep.

8. To set off parenthetical expressions. Parenthetical expressions are statements that provide additional information; however, they interrupt the flow of the sentence.

EXAMPLE: Fluency in a foreign language, as we all know, requires years of instruction and practice.

A comma *should not* be used in the following situations.

1. After *that,* when *that* is used in an introductory clause

EXAMPLE: The student could not believe that he lost points on his laboratory report because of a few spelling mistakes.

2. Between cumulative adjectives, which are adjectives that would not make sense if separated by the word *and* (see Item 4 in preceding list)

The semicolon. The semicolon inserts a stop between two independent clauses not joined by a coordinating conjunction (*and, but, or, nor, for, so,* or *yet*). Each independent clause (one that contains a subject and a verb) could stand alone as a sentence, but the semicolon indicates a closer relationship between the clauses than if they were written as separate sentences.

EXAMPLE: Outstanding student-athletes use their time wisely; this trait makes them highly sought after by employers.

A semicolon is also used to separate items in a series in which the items are already separated by commas.

EXAMPLE: Participating in sports has many advantages. First, you are doing something good for your

health;second, you enjoy the camaraderie of people with a common interest;third, you learn discipline, which helps you make effective use of
your time.

The colon. The colon is used to call attention to the words that follow it.
Some conventional uses of a colon are shown in the following examples.

Dear Sir or Madam:
5:30 P.M.
2:1 (ratio)

In references, to separate the publisher's location and name, as in

Sunderland, MA: Sinauer Associates, Inc.

A colon is often used to set off a list, as in the following example.

EXAMPLE: Catalase activity has been found in the following
vegetables:potatoes, leeks, parsnips, onions, zucchini, carrots, and broccoli.

A colon *should not* be used when the list follows the words *are, consist of,
such as, including,* or *for example.*

EXAMPLE: Catalase activity has been found in vegetables
such as potatoes, leeks, parsnips, onions,
zucchini, carrots, and broccoli.

The period. The period is used to end all sentences except questions
and exclamations. It is also used in some abbreviations, for example, *Mr.,
Ms., Dr., Ph.D., i.e.,* and *e.g.*

Parentheses. Parentheses are used mainly in two situations in scientific writing: to enclose supplemental material and to enclose references for
citations. Use parentheses sparingly because they interrupt the flow of
the sentence.

EXAMPLE: Human error(failure to make the solutions correctly, arithmetic errors, and failure to zero the
spectrophotometer)was the main reason for the
unexpected results.

[Citation-sequence system]   C-fern spores do not germinate
in the dark (1).

[Name-year system]   C-fern spores do not germinate in
                            the dark (Cooke and others, 1987).

The dash. The dash is used to set off material that requires special emphasis. To make a dash on the computer, type two hyphens without a space before, after or in between. In some word processing programs, the two hyphens are automatically converted to a solid dash.

Similar to commas and parentheses, a pair of dashes may be used to set off supplemental material.

EXAMPLE: Human error--failure to make the solutions cor-
           rectly, arithmetic errors, and failure to zero the
           spectrophotometer--was the main reason for the
           unexpected results. (If the word processing pro-
           gram has been set up to convert the two hyphens
           to a solid dash, the sentence looks like this:
           Human error—failure to make…spectropho-
           tometer—was the main reason…)

Similar to a colon, a single dash calls attention to the information that follows it.

EXAMPLE: Catalase activity has been found in many vegeta-
           bles͞potatoes, leeks, parsnips, onions, zucchini,
           carrots, and broccoli.

If an abrupt or dramatic interruption is desired, use a dash. If the writing is more formal or the interruption should be less conspicuous, use one of the other three punctuation marks. However, do not replace a pair of dashes with commas when the material to be set off already contains commas, as in the following example.

EXAMPLE: The instruments that she plays͞oboe, guitar, and
           piano͞are not traditionally used in the marching
           band.

Clarity

The main reason for writing a scientific paper or laboratory report is to communicate information to your peers. If you do not present this information clearly, your readers will not understand what you mean. For student writers whose papers are evaluated by instructor readers, lack of clarity translates into a low grade. Researchers and faculty members, whose reputation depends on the number and quality of publications, simply cannot afford *not* to write clearly, because poorly written papers may be equated with shoddy scientific methods.

There are a number of ways to improve the clarity of your writing. These include eliminating wordiness, redundancy, empty phrases, and ambiguity; reducing complexity; and ensuring good flow between sentences and paragraphs.

Eliminate wordiness. "Wordiness" means using too many words to convey an idea. One form of wordiness is redundancy, using two or more words that mean the same thing (Table 5.2).

Redundancy is easily corrected by eliminating one of the redundant words.

TABLE 5.2 Examples of redundancy

REDUNDANT	REVISED
It is absolutely essential…	It is essential…
mutual cooperation	cooperation
basic fundamental concept	basic concept or fundamental concept
totally unique	unique
The solution was obtained and transferred…	The solution was transferred…

Empty phrases are another source of wordiness. Consider the following two sentences (inspired by Van Alstyne, 1986). Which one would you rather read?

EXAMPLE 1: It is absolutely essential that you use a minimum number of words in view of the fact that your reader has numerous other tasks to complete at the present time.

EXAMPLE 2: Write concisely, because your reader is busy.

Replace empty phrases with a concise alternative (Table 5.3).

A third source of wordiness is **needlessly complex sentences**. Watch out for the sentence constructions illustrated by the following examples.

EXAMPLE 1: *There are* two protein assays *that* are often used in research laboratories.

REVISION: Two protein assays are often used in research laboratories. [Avoid expletives.]

TABLE 5.3 Examples of empty phrases

EMPTY	CONCISE
a downward trend	a decrease
a great deal of	much higher
a majority of	most
accounted for the fact that	because
as a result	so, therefore
as a result of	because
as soon as	when
at which time	when
at all times	always
at a much greater rate than	faster
at the present time, at this time	now, currently
based on the fact that	because
brief in duration	short, quick
by means of	by
came to the conclusion	concluded
despite the fact that, in spite of the fact that	although, though
due to the fact that, in view of the fact that	because
for this reason	so

EXAMPLE 2:   *It is interesting to note that* some enzymes are stable at temperatures above 60°C.

REVISION: Some enzymes are stable at temperatures above 60°C. [Avoid unnecessary introductions.]

EXAMPLE 3:   *The analyses were done on* the recombinant DNA to determine which piece of foreign DNA was inserted into the vector.

REVISION: The recombinant DNA was analyzed to determine which piece of foreign DNA was inserted into the vector. [Make *DNA*, not the *analyses*, the subject of the sentence.]

TABLE 5.3 Examples of empty phrases

EMPTY	CONCISE
in fact	*omit*
functions to, serves to	*omit*
degree of	higher, more
in a manner similar to	like
in the amount of	of
in the vicinity of	near, around
is dependent upon	depends on
is situated in	is in
it is interesting to note that, it is worth pointing out that	*omit these kinds of unnecessary introductions*
it is recommended	I (we) recommend
on account of	because, due to
prior to	before
provided that	if
referred to as	called
so as to	to
through the use of	by, with
with regard to	on, about
with the exception of	except
with the result that	so that

EXAMPLE 4: *We make the recommendation* that micropipettors be used to measure volumes less than 1 mL.

REVISION: We recommend that micropipettors be used to measure volumes less than 1 mL. [Replace sluggish noun phrases with verb phrases.]

EXAMPLE 5: These assays alone cannot tell what the protein concentration of a substance is.

REVISION: These assays alone cannot determine the protein concentration of a substance. [Replace colloquial expressions with precise alternatives.]

Eliminate ambiguity.   Avoid vague use of *this, that,* and *which* when referring to topics mentioned previously. What does the writer mean by *this* and
which in the following example?

> EXAMPLE: The data show that the longer the enzyme
> was exposed to the salt solution, the lower
> the enzyme activity in the assay. *This* means
> that the salt changes the conformation of the
> enzyme, *which* makes it less reactive with
> the substrate.

The subsequent revision eliminates the ambiguity.

> REVISION: The data show that the longer the enzyme was
> exposed to the salt solution, the lower the enzyme
> activity in the assay. Exposure to the salt solution
> may change the conformation of the enzyme,
> resulting in lower enzyme-substrate activity.

Another source of ambiguity is when a pronoun (him, her, it, he, she, its)
could refer to two possible antecedents. In the following example, can
you determine what the writer meant by *it*?

> EXAMPLE: With time, salt changes the conformation of
> the enzyme, which makes *it* less reactive with
> the substrate.

If there is any doubt about what *it* refers to, replace *it* with the appropriate noun phrase.

> REVISION: With time, salt changes the conformation of the
> enzyme, so that the enzyme can no longer react
> with its substrate.

Use connecting words and repetition to improve flow.   A smooth transition from one sentence to the next is essential for reader comprehension.
Consider the following example:

> EXAMPLE: Catalase is an enzyme that breaks down hydro-
> gen peroxide in both plant and animal cells. Low
> or high temperature can lower the rate at which
> the catalase can react with the hydrogen perox-
> ide. In optimal conditions, the enzyme functions
> at a rate that will prevent any substantial buildup
> of the toxin. If the temperature is too low, the rate
> will be too slow, but high temperatures lead to
> the denaturation of the enzyme.

Where is the writer going with this paragraph? The sentences do not seem to flow, because there is no guidance from the writer on how one sentence is related to the next. To improve flow, use connecting words such as *however, thus, although, in contrast, similarly, on the other hand, in addition to,* and *furthermore.* Repetition of a key word from the previous sentence also helps the reader make connections between sentences.

> REVISION: Catalase is an enzyme that breaks down hydro-
> gen peroxide in both plant and animal cells. One
> of the factors that affects the rate *of this reaction* is
> temperature. At optimal *temperatures,* the rate is
> sufficient to prevent substantial buildup of the
> toxic hydrogen peroxide. If the temperature is
> too low, *however,* the rate will be too slow, and
> hydrogen peroxide *accumulates* in the cell. *On the
> other hand,* high temperatures may denature the
> enzyme.

Grammar

Grammar refers to the rules that deal with the form and structure of words and their arrangement in sentences. This section describes three of the most common offenses. See Hacker (1997), Lunsford (2001), or Lunsford and Connors (1995, 1999) for a more comprehensive treatment of the subject.

Make subjects and verbs agree. We learn early on in our formal educa-tion to make the verb agree with the subject. Most of us know that *the sample was…*, but that *the samples were…* Most errors with subject-verb agreement occur when there are words *between* the subject and the verb, as in the following example.

> EXAMPLE: *The samples* in the assay *were* [not was] incubated
> at 37°C for 10 min.

When you write complex sentences, ask yourself what is the subject of the sentence. Look for the verb that goes with that subject, and then men-tally remove the words in between. Make the subject and its verb agree.

A second situation in which subject–verb agreement becomes confusing is when there are two subjects joined by *and,* as in the following example.

> EXAMPLE: An enzyme's amino acid *sequence and* its three-
> dimensional *structure make* [not makes] the
> enzyme-substrate relationship unique.

Compound subjects joined by *and* are almost always plural.

A third situation involving subject–verb agreement is that when numbers are used in conjunction with units, the quantity is considered to be singular, not plural.

> EXAMPLE: To extract the enzyme, 50 g of potatoes *was* [not were] homogenized with 50 mL of cold, distilled water.

Write in complete sentences.   A complete sentence consists of a subject and a verb. If the sentence starts with a subordinate word or words such as *after, although, because, before, but, if, so that, that, though, unless, until, when, where, who,* or *which,* however, another clause must complete the sentence.

> EXAMPLE 1:   High temperatures destroy the three-dimensional structure of enzymes. Thus changing the effectiveness of the enzymes. [The second "sentence" is a fragment.]

> REVISION 1:   High temperatures destroy the three-dimensional structure of enzymes, thus changing their effectiveness. [Combine the fragment with the previous sentence, changing punctuation as needed.]

> EXAMPLE 2:   The standard curve for the Biuret assay was used to determine the protein concentration of the serial dilutions of the egg white. Although only those dilutions whose protein concentrations fell within the sensitivity range of the assay were multiplied by the dilution factor to give the original concentration of the egg white. [The second "sentence" is a fragment.]

> REVISION 2:   The standard curve for the Biuret assay was used to determine the protein concentration of the serial dilutions of the egg white. Only those dilutions whose protein concentrations fell within the sensitivity range of the assay were multiplied by the dilution factor to give the original concentration of the egg white. [Delete the subordinate word(s) to make a complete sentence.]

Revise run-on sentences. Run-on sentences consist of two or more independent clauses joined without proper punctuation. Each independent clause could stand alone as a complete sentence. Run-on sentences are common in first drafts, where your main objective is to get your ideas down on paper (or electronic media, if you use a computer). When you revise your first draft, however, use one of the following strategies to revise run-on sentences:

- Insert a comma and a coordinating conjunction (*and, but, or, nor, for, so,* or *yet*).
- Use a semicolon or possibly a colon.
- Make two separate sentences.
- Rewrite the sentence.

EXAMPLE 1: The class data for the Bradford method were scattered, those for the Biuret method were closer.

REVISION 1A: The class data for the Bradford method were scattered, but those for the Biuret method were closer. [Use a coordinating conjunction.]

REVISION 1B: The class data for the Bradford method were scattered; those for the Biuret method were closer. [Use a semicolon.]

EXAMPLE 2: The readings from the spectrophotometer should show a correlation between protein concentration and absorbance, this is Beer's law, which relates absorbance to the path length of light along with molar concentration of a solute and the molar coefficient. [Fused sentence]

REVISION 2A: The readings from the spectrophotometer should show a correlation between protein concentration and absorbance; this is Beer's law, which relates absorbance to the path length of light along with molar concentration of a solute and the molar coefficient. [Use a semicolon to separate the two clauses.]

REVISION 2B: The readings from the spectrophotometer
should show a correlation between protein
concentration and absorbance. This relationship is described by Beer's law, which relates
absorbance to the path length of light along
with molar concentration of a solute and
the molar coefficient. [Make two separate
sentences.]

EXAMPLE 3:   An increase in enzyme concentration
increased the reaction rate as did an increase
in substrate concentration, so the concentrations of the molecules have an influence on
how the enzyme reacts.

REVISION 3A: As enzyme concentration and substrate concentration increased, so did the reaction rate.
[Rewrite the sentence. The second half of the
original sentence was deleted because it is
deadwood.]

REVISION 3B: Enzyme and substrate concentration influence enzyme reaction rate: an increase in
enzyme or substrate concentration increased
reaction rate. [Use a colon.]

Word usage

When you write the right words in the right situations, readers have
confidence in your work. Use a standard dictionary whenever you are
not sure about word usage. Consult your textbook and laboratory exercise for proper spelling and usage of technical terms. The following
word pairs are frequently confused in students' biology papers.

absorbance, absorbency, observance   *Absorbance* is how much
light a solution absorbs; absorbance is measured with a spectrophotometer. *According to Beer's law, absorbance is proportional to
concentration. Absorbency* is how much moisture a diaper or paper
towel can hold. *Brand A paper towels show greater absorbency than
Brand B paper towels. Observance* is the act of observing. *Government
offices are closed today in observance of Independence Day.*

affect, effect   *Affect* is a verb that means "to influence." Affect is never used as a noun. *Temperature affects enzyme activity. Effect* can be used either as a noun or a verb. When used as a noun, *effect* means "result." *We studied the effect of temperature on enzyme activity.* When used as a verb, effect means "to cause." *High temperature effected a change in enzyme conformation, which destroyed enzyme activity.*

alga, algae   See Plurals.

amount, number   Use *amount* when the quantity cannot be counted. *The reaction rate depends on the amount of enzyme in the solution.* Use *number* if you can count individual pieces. *The reaction rate depends on the number of enzyme molecules in the solution.*

analysis, analyses   See Plurals.

bacterium, bacteria   See Plurals.

bind, bond   *Bind* is a verb meaning "to link." *The active site is the region of an enzyme where a substrate binds. Bond* is a noun that refers to the chemical linkage between atoms. *Proteins consist of amino acids joined by peptide bonds. Bond* used as a verb means "to stick together." *This 5-minute epoxy glue can be used to bond hard plastic.*

complementary, complimentary   *Complementary* means "something needed to complete;" matching. *The DNA double helix consists of complementary base pairs: A always pairs with T, and G with C. Complimentary* means "given free as a courtesy." *The brochures at the Visitor's Center are complimentary.*

confirmation, conformation   *Confirmation* means "verification." *I received confirmation from the postal service that my package had arrived. Conformation* is the three-dimensional structure of a macromolecule. *Noncovalent bonds help maintain a protein's stable conformation.*

continual, continuous   *Continual* means "going on repeatedly and frequently over a period of time." *The continual chatter of a group of inconsiderate students during the lecture annoyed me.*

Continuous means "going on without interruption over a period of time." *The bacteria were grown in L-broth continuously for 48 hr.*

create, prepare, produce   *Create* is to cause to come into existence. *The artist used wood and plastic to create this sculpture. Prepare* means "to make ready." *The protein standards were prepared from a 50 mg/mL stock solution. Produce* means to make or manufacture. *The reaction between hydrogen peroxide and catalase produces water and oxygen.*

datum, data   See Plurals.

different, differing   *Different* is an adjective that means "not alike." An adjective modifies a noun. *Different concentrations of bovine serum albumin were prepared. Differing* is the intransitive tense of "to differ," a verb that means "to vary." It is incorrect to replace the word *different* with *differing* in the preceding example, because *differing* implies that a single concentration changes depending on time or circumstance. This situation is highly unlikely with bovine serum albumin, which is quite stable under laboratory conditions! An acceptable use of *differing* is shown in the following example. *Bovine serum albumin solutions, differing in their protein content, were prepared.*

effect, affect   See affect, effect.

fewer, less   Use *fewer* when the quantity can be counted. *The reaction rate was lower, because there were fewer collisions between enzyme and substrate molecules.* Use *less* when the quantity cannot be counted. *The weight of this sample was less than I expected.*

formula, formulas, formulae   See Plurals.

hypothesis, hypotheses   See Plurals.

its, it's   *Its* is a possessive pronoun meaning "belonging to it." *The Bradford assay is preferred because of its greater sensitivity. It's* is a contraction of "it is." *The Bradford assay is preferred because it's more sensitive.* (*Note:* Contractions should not be used in formal writing.)

less, fewer   See fewer, less

TABLE 5.4 Singular and plural of some common biological words

SINGULAR	PLURAL
alga	algae
analysis	analyses
bacterium	bacteria
criterion	criteria
datum (rarely used)	data
formula	formulas, formulae
hypothesis	hypotheses
index	indexes, indices
medium	media
phenomenon	phenomena
ratio	ratios

media, medium   See Plurals.

observance   See absorbance, absorbency, observance.

phenomenon, phenomena   See Plurals.

Plurals   The plural and singular forms of some words used in biology are given in Table 5.4. A common mistake with these words is failure to make the subject and verb agree. Some disciplines treat *data* as singular, but scientists and engineers subscribe to the strict interpretation that *data* is plural. The data *show*… [not *shows*] is correct.

prepare   See create, prepare, produce.

produce   See create, prepare, produce.

ratio, ration   *Ratio* is a proportion or quotient. *The ratio of protein in the final dilution was 1:5. Ration* is a fixed portion, often referring to food. *The Red Cross distributed rations to the refugees.*

strain, strand   A *strain* is a line of individuals of a certain species, usually distinguished by some special characteristic.

The lacI⁻ strain of E. coli produces a nonfunctional repressor protein. A strand *is a ropelike length of something. The strands of DNA are held together with hydrogen bonds.*

that, which   Use *that* with restrictive clauses. A restrictive clause limits the reference to a certain group. Use *which* with nonrestrictive clauses. A nonrestrictive clause does not limit the reference, but rather provides additional information. Commas are used to set off nonrestrictive clauses but not restrictive clauses. Consider the following examples:

EXAMPLE 1:   The Bradford assay, which is one method for measuring protein concentration, requires only a small amount of sample. [*Which* begins a phrase that provides additional information, but is not essential to make a complete sentence.]

EXAMPLE 2:   Enzyme activity decreased significantly, which suggests that the enzyme was denatured at 50°C. [*Which* refers to the entire phrase *Enzyme activity decreased significantly*, not to any specific element.]

EXAMPLE 3:   The samples that had high absorbance readings were diluted. [*That* refers specifically to *The samples*.]

than, then   *Than* is an expression used to compare two things. *Collisions between molecules occur more frequently at high temperatures than at low temperatures.* Then means "next in time." *First 1 mL of protein sample was added to the text tube. Then 4 mL of Biuret reagent was added.*

various, varying   *Various* is an adjective that means "different." *Various hypotheses were proposed to explain the observations. Varying is a verb that means "changing." Varying the substrate concentration while keeping the enzyme concentration constant allows you to determine the effect of substrate concentration on enzyme activity.* Analogous to *different, differing,* replacing the word *various* with *varying* in the preceding example changes the meaning of the sentence. *Varying* implies that a single hypothesis changes depending on time or circumstance. *Various* implies that different hypotheses were proposed.

Spelling

Spell checkers in word processing programs are so easy to use that there is really no excuse for *not* using them. Just remember that spell checkers may not know scientific terminology, so consult your textbook or laboratory manual for correct spelling. In some cases, the spell checker may even try to get you to change a properly used scientific word to an inappropriate word that happens to be in its database (for example, *absorbance* to *absorbency*).

The following poem is an example of how indiscriminate use of the spell checker can produce garbage:

Wrest a Spell

Eye halve a spelling chequer
It came with my pea sea
It plainly marques four my revue
Miss steaks eye kin knot sea.

Eye strike a key and type a word
And weight four it two say
Weather eye am wrong oar write
It shows me strait a weigh.

As soon as a mist ache is maid
It nose bee fore two long
And eye can put the error rite
Its rare lea ever wrong.

Eye have run this poem threw it
I am shore your pleased two no
Its letter perfect awl the weigh
My chequer tolled me sew.

— Sauce unknown

Spell checkers will also not catch mistakes of usage, for example *form* if you really meant *from*. Print out your document and proofread the hard copy carefully.

Global Revision

Organization. Is your document formatted according to "Instructions to Authors"? Is your scientific paper divided into sections? Does each section have the right content? The "Revision Checklist" and "Laboratory Report Mistakes" tables can alert you to potential problem areas.

Content. Is your paper focussed on the subject? Did you provide all of the necessary supporting materials (tables and figures) to support your conclusions? Did you answer all of the questions in the laboratory exercise? Use the "Revision Checklist" and "Laboratory Report Mistakes" key.

Style. Read your paper aloud. Listen for awkward sentence structure and illogical reasoning. Avoid wordiness and flowery language that might make your meaning ambiguous. Use definite and specific sentences. Keep related words together. Consider the following sentence taken from an English-language newspaper in Japan: "A committee was formed to examine brain death in the Prime Minister's office." Although brain death in the Prime Minister's office may be a political reality, what was really intended was "A committee was formed in the Prime Minister's office to examine brain death."

Focus on the following to achieve a clear and readable style.

1. Paragraphs
   - Each paragraph focuses on one topic.
   - First sentence introduces the topic.
   - Subsequent sentences support the topic sentence.
   - Connecting phrases are used to achieve good flow between sentences.

2. Sentences
   - Eliminate wordiness.
   - Vary sentence structure and length to avoid monotony.
   - Use punctuation correctly.
   - Use passive and active voice appropriately.
   - Use past and present tense appropriately.

3. Words and phrases
   - Make them concise and descriptive.
   - Use scientific words when appropriate. Define terms that might be unfamiliar to your audience. Avoid jargon and anthropomorphism.
   - Watch out for commonly confused word pairs such as *effect* and *affect, it's* and *its, then* and *than,* and others.
   - Avoid clichés, slang, and sexist language.
   - Do not use contractions in formal writing.

TITLE
☐ Descriptive and concise

AUTHORS
☐ Writer is first author, lab partner(s) second (third, etc.) author

ABSTRACT
☐ Contains purpose of experiment
☐ Contains brief description of methods
☐ Contains results
☐ Contains conclusions

INTRODUCTION
☐ Contains background information from the literature (primary references)
☐ Selected references are directly relevant to your experiment.
☐ Citation format is correct.
☐ Citations are paraphrased. Quotation marks are not used.
☐ Objectives of experiment are clearly stated.

MATERIALS AND METHODS
☐ Materials are not listed separately.
☐ Written in paragraph form (not listed like steps in a recipe)
☐ Written in past tense
☐ Contains all relevant information to enable the reader to repeat the experiment. This includes volumes, temperatures, wavelengths, and so on.

RESULTS
☐ Figures and tables are present. Figure captions go below figure; table captions go above table
☐ Text is present. Reference is made to each table and figure, and the results are described in words.

Revision Checklist

☐ Figure and table titles are informative and can be understood apart from the text.

☐ No explanation is given for the results.

DISCUSSION

☐ Results are briefly restated.

☐ Explanations for results are given; implications are considered.

☐ Errors and inconsistencies are pointed out.

REFERENCES

☐ References consist mostly of primary journal articles, not textbooks or Internet sources.

☐ Reference format is correct and complete.

☐ All references have been cited in the text. All citations in the text have been included in the References section.

REVISION

☐ Conventions in biology

☐ Numbers

☐ Standard abbreviations

☐ Punctuation

☐ Clarity

☐ Grammar

☐ Word usage

☐ Spelling

☐ Global revision

☐ All questions from the laboratory exercise have been answered

Revision Checklist

Chapter 6

A "Good" Sample Student Laboratory Report

The laboratory report in this chapter was written by Lynne Waldman during her first year at Bucknell University, in an introductory course for biology majors. Lynne and her lab partners designed and carried out an original project in which they investigated the effect of a fungus on the growth of bean, pea, and corn plants.

Lynne's report has many of the characteristics of a well-written scientific paper. When you look over her presentation, notice the style and tone of her writing, as well as the format of the paper. The comments and annotations alert you to important points to keep in mind when you write your laboratory report.

The presentation here has been typeset to fit this book and to accommodate the annotations. Your report should be formatted to fit standard $8\frac{1}{2} \times 11$ inch paper. Unless you are instructed otherwise, use a serif type (Times Roman is standard), double space, and leave *at least* 1 inch of margin all around.

For details on how to format documents in Microsoft® Word, see Appendix I, "Word Processing Basics."

The Effects of the Fungus *Phytophthora infestans* on Bean, Pea, and Corn Plants

Lynne Waldman. Partner One, Partner Two

Abstract

Phytophthora infestans is a fast-spreading, parasitic fungus that caused the infamous potato blight by devastating Ireland's crops in the 1840s. *P. infestans* also causes late blight in tomato plants, a relative of the potato. In this experiment, the effects of *P. infestans* on *Phaseolus* variety long bush bean, *Zea mays* (corn), and *Pisum sativum* (pea) were studied. The soil surrounding the roots of 18-day old plants was injected with *P. infestans* cultured in an L-broth medium. Plant height, number of leaves, and leaf angle were measured for each plant during the next 8 days. Chlorophyll assays were performed prior to exposure, and on the eighth day after exposure to the fungus. The plants were also examined for black or brown leaf spots characteristic of late blight infections. The results showed that *P. infestans* had no apparent effect on the bean, corn, and pea plants. One reason for this may be that there were no fungus zoospores in the L-broth medium. More probably, however, *P. infestans* may be a species-specific pathogen that cannot infect bean, corn, or pea plants.

Introduction

Originating in Peruvian-Bolivian Andes, the potato *(Solanum tuberosum)* is one of the world's four most important food crops (along with wheat, rice, and corn). Cultivation of potatoes began in South America over 1,800 years ago, and through the Spanish conquistadors, the tuber was introduced into Europe in the second half of the 1600s. By the beginning of the 18th century, the potato was widely grown in Ireland, and the country's economy heavily relied on the potato crop. In the middle of the 19th century, Ireland's potato crop suffered wide-

---

Margin annotations (left column):

- Informative title
- Author's name first, followed by lab partners' names in alphabetical order
- Sections of report are clearly labeled.
- Background information is present.
- No references are used in Abstract.
- Purpose of current experiment is given.
- Short description of methods
- Results are given.
- No references are made to figures.
- Short explanation of results is given.
- Abstract is a maximum of 250 words long.
- Background information
- Latin names are italicized.

spread late blight disease caused by *Phytophthora infestans,* a species of pathogenic plant fungus. Failure of the potato crop because of late blight resulted in the Irish potato famine. The famine led to widespread starvation and the death of about a million Irish.

The potato continues to be one of the world's main food crops. However, *P. infestans* has reemerged in a chemical-resistant form in the United States, Canada, Mexico, and Europe (McElreath, 1994). Late blight caused by the new strains is costing growers worldwide about $3 billion annually. The need to apply chemical fungicides eight to ten times a season further increases the cost to the grower (Stanley, 1994 and Stanley, 1997). *P. infestans* is thus an economically important pathogen.

P. infestans, which can destroy a potato crop in the field or in storage, thrives in warm, damp weather. The parasitic fungus causes black or purple lesions on a potato plant's stem and leaves. As a result of infection by this fungus, the plant is unable to photosynthesize, develops a slimy rot, and dies. *P. infestans* similarly infects the tomato plant *(Lycopersicon esculentum)* (Anonymous, 1994).

The purpose of the present experiment was to determine the effects of *P. infestans* on plant height, number of leaves, leaf angle, and chlorophyll content of three agriculturally important plants: *Phaseolus* variety long bush beans, *Zea mays* (corn), and *Pisum sativum* (peas). Symptoms of fungal infection were assumed to be similar to that in potatoes.

Materials and Methods

Phaseolus variety long bush bean, *Zea mays* (corn) and *Pisum sativum* (pea) seeds were soaked overnight in tap water. Fifteen randomly chosen seeds of each species were planted 1 cm beneath the surface in three separate trays containing 10 cm of potting soil. Another set of trays, which was to be the control group, was prepared in the same fashion.

Side annotations:

Proper citation format is used (Name-Year system).

No quotation marks are used. Citations are paraphrased.

"Anonymous" is used when no author is given in newspaper or magazine articles.

Purpose of experiment is clearly stated.

M&M are always written in past tense.

Sufficient detail is given to allow the reader to repeat the experiment.

All the experimental plants were placed in one fume hood, and all the control plants were placed in relative positions in another fume hood in the same room. The plants were exposed to the ambient light intensity in the hood (153 fc) and air current 24 hrs a day, and were watered lightly daily. The plants were allowed to germinate and grow for 18 days.

Phytophthora infestans on potato dextrose agar was obtained from Carolina Biological Supply House. At day 10 of the plant growth regime, pieces of agar on which the fungus was growing were transferred to L-broth. L-broth consisted of 5 g yeast extract, 10 g tryptone, 1 g dextrose, and 10 g NaCl dissolved in distilled water, and adjusted to pH 7.1, to make 1 L of medium. The medium was sterilized before adding the fungal culture. After 4 days in L-broth, 6 mL of the fungal culture were injected into the soil around the roots of each 18-day old plant. 6 mL of L-broth without *P. infestans* was injected into the soil of the control plants. All plants were then allowed to grow for another 8 days.

Every other day after treatment with *P. infestans*, plant height and number of leaves were measured for both the control and the experimental plants. Plant height was measured from the soil to the apical meristem of the plant. Leaf angle (as shown in Figure 1) of the largest, lowest leaf on each plant was measured three times, once prior to injection, once 4 days after injection, and once 8 days after injection. Leaf angle was measured in order to

Figures that explain the methodology may be included in M&M section.

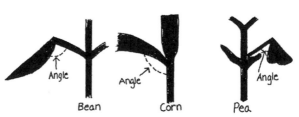

Figure 1. Leaf angle as measured in bean, corn, and pea plants

determine if *P. infestans* causes wilting in the three plant species. In addition, the plant was examined visually for the presence of any leaf spots.

Chlorophyll assays were performed on one plant from each tray prior to injection and on the eighth day after injection. For each chlorophyll assay, the leaves of the plant were removed from the stem. For each 0.1 g of leaves, 6.0 mL of 100% methanol were used. The leaves were thoroughly ground in half of the methanol with a pestle in a mortar. The leaves were ground again after the rest of the methanol was added. Extraction of the chlorophyll was allowed to proceed for 45 min at room tempera-ture. Then the suspension was gravity filtered through filter paper to remove the leaf parts. The absorbance of the filtrate was measured with a Spectronic 20 spectrophotometer at 652 nm and 665.2 nm. The absorbance values were converted to relative chlorophyll units using the following equa-tion derived by Porra and colleagues (1989):

Total chlorophyll (a and b) = Dilution factor × $[22.12 A_{652\ nm} + 2.71A_{665.2\ nm}$ (mg/L)] × Volume of solvent (L) / Weight of solvent (mg)

Subscripts are made properly.

Results
P. infestans-treated plants and the control plants had similar growth patterns (Figure 2). Both the experi-mental and control pea and corn plants grew at a

Results section must contain a text in which the author presents each figure and table to the reader.

Reference is made to the next figure in the sequence, and the important results are described.

Figure is large enough to read key and axes easily.

Intervals on axes are regular and proportional.

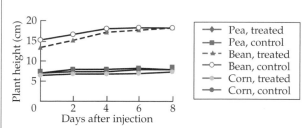

Figure 2. Average height of control and experimental plants in the period after injection with *P. infestans*

Points and lines on curve are dark and can be easily distinguished from each other.

Units are given in parentheses after the axis legend (where applicable).

Figure caption is located below figure.

Figure title is not simply a repeat of "*y*-axis legend vs. *x*-axis legend."

Figure number sequence is correct.

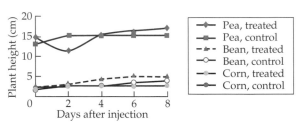

Figure 3. Average number of leaves of control and experimental plants in the period after injection with *P. infestans*

Reference is made to the next figure and the important information is described.

Symbols (such as °) are typed using word processing software.

constant, but very slow rate over the eight day test period. The control bean plants were taller on average than the experimental bean plants throughout most of the experiment. Both groups showed the same growth pattern, however, with rapid growth occurring from day 18 to 24 (0 to 4 days after injection), followed by slower growth to the end of the experiment.

Figure 3 shows that, as plant height increased, the average number of leaves on all of the plants also increased over the measurement period. There is an uncharacteristic decrease in the number of leaves of pea plants treated with *P. infestans* from day 18 to 20 (0 to 2 days after injection), but this is probably due to counting error.

There was a general decline in average leaf angle of all the plants over the first four days after injection with *P. infestans* (Figure 4). The plants did not follow this pattern over the second half of the experiment, however. The leaf angle of the experimental bean group increased by 28°, while that of the control bean group only increased by about 3°. The leaf angle of the control pea plants increased significantly (33°), while that of the experimental pea plants decreased 4°. The leaf angle of the corn control group decreased 0.5°, while that of the corn experimental group showed a much sharper decline of 24°.

There was also no difference between the experimental and control groups with regard to chloro-

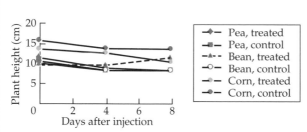

Figure 4. Average leaf angle of control and experimental plants in the period after injection with *P. infestans*

phyll content. Table 1 shows that there was a slight increase in chlorophyll content from day 18 to 26 (0 to 8 days after injection) in the corn plants. For the bean group, there was a large decrease in chlorophyll content, 0.1 relative chlorophyll units, which did not seem to agree with the general appearance of the plants. There may have been some error when this assay was carried out. There was little change in chlorophyll content for the pea group.

Finally, there was no evidence of any brown or black leaf spots symptomatic of *P. infestans* infection.

Table is numbered properly.

Table 1. Chlorophyll content of corn, bean, and pea plants prior to infection and 8 days after infection

Table caption is placed above the table.

Plant	Relative chlorophyll units		Change in chlorophyll content (relative units)
	Day 0	Day 8	
Corn, treated	9.036×10^{-4}	9.383×10^{-4}	$+3.45 \times 10^{-5}$
Corn, control	9.270×10^{-4}	8.963×10^{-4}	$+3.34 \times 10^{-5}$
Bean, treated	1.034×10^{-1}	1.2×10^{-3}	-1.022×10^{-1}
Bean, control	1.7×10^{-3}	1.6×10^{-3}	-1×10^{-4}
Pea, treated	1.3×10^{-3}	1.7×10^{-3}	$+4 \times 10^{-4}$
Pea, control	1.2×10^{-3}	1.2×10^{-3}	0.0000

Scientific notation is used correctly.

It is preferable to leave extra space at the bottom of the page, rather than to split a table across a page.

Discussion

Results are summarized or briefly restated in the Discussion section.

The data for plant height, leaf angle, number of leaves, and chlorophyll content show that *P. infestans* did not have an effect on *Zea mays, Pisum sativum,* and *Phaseolus.* Symptoms of infection are the presence of brown or black spots (areas of dead tissue) on leaves and stems, and, as the infection spreads, the entire plant becomes covered with a cottony film (Stanley, 1994). None of the experimental plants exhibited these symptoms.

Explanations for results are given.

There may be several reasons why *P. infestans* did not affect the plants in this study. One reason is that the L-broth culture of *P. infestans* may not have contained zoospores of the fungus. Zoospores are motile spores that can penetrate the host plant through the leaves and soft shoots, or through the roots (Stanley, 1994). Zoospores are usually produced in wet, warm weather conditions (Ingold and Hudson, 1993). If the L-broth culture did not contain any zoospores, or if the soil around the plants was not sufficiently saturated to stimulate production of zoospores, then these conditions may have prevented *P. infestans* from attacking the roots and shoots of the plants.

References are used to support explanations.

Ways to test explanations may be offered.

In order to determine if the problem was lack of zoospores, first the L-broth culture could be examined microscopically for presence of zoospores. Second, the *P. infestans* plants could be watered with different quantities of water to determine if the fungus requires wetter soil for zoospore production and motility.

Frequent use of references is made to support explanations. Whenever possible, use primary references (journal articles, conference proceedings, collections of primary articles in a book). Avoid textbooks (secondary references) and Internet sources (may be unreliable).

Another reason why *P. infestans* may not have affected the plants is that this species of fungus may be specific to potato *(Solanum tuberosum)* and tomato *(Lycopersicon esculentum)* plants (Stanley, 1994), which both belong to the nightshade family (Solanaceae). In contrast, corn belongs to the grass family (Gramineae), and peas and beans are legumes (Leguminosae). It may be that these plant families are not susceptible to *P. infestans*, which has a very limited host range (Stanley, 1994). Nonsusceptible plants have been shown to have defense mechanisms that prevent *P. infestans* from infecting them (Gallegly, 1995).

Further research is required to determine if *P. infestans* really cannot infect corn, pea, and bean plants. Goth and Keane (1997) developed a test to measure resistance of potato and tomato varieties to original and new strains of *P. infestans*. Their experiments involved exposing the experimental plants' leaves directly to the fungus, and this method could perhaps be tested on corn, pea, and bean leaves as well.

When Name-Year citation format is used, authors are listed alphabetically in the References section.

References

[Anonymous]. 1994. Brave New Potato. Discover 15(10): 18–20.

Gallegly ME. 1995. New criteria for classifying Phytophthora and critique of existing approaches. In: Phytophthora: Its Biology, Taxonomy, Ecology, and Pathology (Erwin DC, Bartnicki-Garcia S, Tsao PH, editors.) St. Paul: The American Phytopathological Society. pp. 167–172.

Goth RW, Keane J. 1997. A detached-leaf method to evaluate late blight resistance in potato and tomato. American Potato Journal 74(5): 347–352.

Ingold CT, Hudson HJ. 1993. The Biology of Fungi, 6th ed. London: Chapman and Hall.

McElreath, Linda R. 1994. One potato, two potato. Agricultural Research 42(5): 2–3.

Porra RJ., Thompson WA, Kriedemann PE. 1989. Determination of accurate extinction coefficients and simultaneous equations for assaying chlorophylls a and b extracted with four different solvents: verification of the concentration of chlorophyll standards by atomic absorption spectroscopy. Biochimica et Biophysica Acta 975: 384–394.

Stanley D. 1994. What was around comes around. Agricultural Research 42(5): 4–8.

Stanley D. 1997. Potatoes. Agricultural Research 45(5): 10–14.

Most references should be primary journal articles or articles in books. Textbooks and articles in newspapers and magazines are secondary references (less preferable).

List all authors (up to 10; then list first 10 followed by "et al." or "and others.")

See the tabbed pages in Chapter 4 for examples of how to reference journal articles, articles in books, and books.

List the starting and ending pages of the article, not just the page(s) you extracted information from.

All citations must have a corresponding reference.

All references must be cited in the text.

Laboratory Report Mistakes

The following table lists some common mistakes made by students writing biology lab reports. When you proofread the first draft of your report, look over this list and be on the lookout for these kinds of mistakes.

Your instructor may also use this key to save time grading lab reports without sacrificing the quality of the feedback. For example, the number "18" written in the left margin means that you failed to include enough detail about the procedure in the Materials and Methods section. If your instructor uses this system, be sure to refer to the key for an explanation of your instructor's comments.

Although some of these mistakes may not affect the content of your paper, they do affect your credibility as a scientist; careless writing may be equated with careless science.

MISTAKE	EXPLANATION
lc	Use lowercase letter
CAP or uc	Use capital (uppercase) letter
∧	Insert text
⌐	Leave space between the two characters
∩∪	Close up
¶	Start a new paragraph
Page break	End the current page; move subsequent text to next page. (In Microsoft® Word and WordPerfect®, press Ctrl + Enter where you want to end the page.)
agr	Subject and verb do not agree.
wc	Word choice. Word used is not appropriate for the situation.
......	A dotted underline means "stet," or "let original text stand." The correction was made in error.
1	Word usage incorrect. Look it up in the dictionary. Examples:
a.	absorbance (how much light is absorbed) vs. absorbency (how much moisture a diaper or paper towel can hold)
b.	to bind (what atoms and molecules do) vs. to bond (what glue does)
c.	complementary (DNA strands) vs. complimentary (given free as a courtesy)
d.	data = plural of "datum"; e.g. "These data show..." *not* "This data shows..."

e.	effect (noun) vs. effect (verb meaning "to cause") vs. affect (verb meaning "to influence")
f.	conformation (3-D shape of a protein) vs. confirmation (verification)
i.	it's (it is) vs. its (belonging to "it")
m.	amount (cannot count individual pieces) vs. number (can count individual pieces)
r.	ratio (proportion or quotient) vs. ration (a fixed portion of food)
s.	strain of bacteria vs. strand of DNA
t.	then (next in time) vs. than (an expression for comparing two things)
v.	varying (changing) vs. various (different)
2	Latin names should be *italicized.*
3	Write in passive voice. Shift the emphasis from yourself to the subject of the action.
4	Punctuation usage is incorrect.
5	Sentence structure is awkward or sentence is incomplete.
6	Symbols should not be handwritten. Use the symbol font or **Insert	Symbol** on the menu bar.
7	Superscript/Subscript. Superscript: For 10^{-5}, for example, this is done by first typing "10-5"; then highlight "-5" and either click on the Superscript icon in the toolbar or select **Format	Font** from the menu bar, and click **Superscript**. Subscript: For K_m, for example, this is done by first typing "*Km*"; then highlight "*m*" and either click on the Subscript icon in the toolbar or select **Format	Font** from the menu bar, and click **Subscript**.
8	Essential Abstract content is missing. Abstract must contain purpose, brief description of methods, results, and possibly an explanation for the results.
9	Too much detail for Abstract.
10	Introduction needs references to provide background.
11	Introduction needs purpose of the current experiment.
12	Wrong citation format. Use either name-year or citation-sequence format. See Chapter 4 for specific examples.
13	Do not use quotation marks in scientific writing. Paraphrase.

Laboratory Report Mistakes

14	Write Materials and Methods in paragraph form. Do not list steps as in a recipe.
15	Do not list materials separately in Materials and Methods section, unless the source is noncommercial or critical for the outcome of the experiment.
16	Time frame does not affect the outcome.
17	Do not explain laboratory techniques that are common knowledge for your audience.
18	Essential details are missing in the Materials and Methods section. Provide enough detail to enable the reader to repeat the experiment.
19	Do not include raw data.
20	Results section always needs a text.
21	Describe *each* table and *each* figure individually and sequentially.
22	Refer to each figure/table by number as you describe the results.
23	A table is not needed when the figure(s) shows the same data.
24	Each figure/table must have a number and a title.
25	The figure/table title must be self-explanatory without reference to the text.
26	Uninformative figure title. Do not use "*y*-axis legend vs. *x*-axis legend" as a title. See Chapter 4 for examples.
27	Figure caption goes *below* figure; table caption goes *above* table.
28	No title is needed above the figure. When you make the figure in Excel, leave the "Chart title" space blank.
29	Intervals on axis are not spaced proportionally. Choose "XY Scatter" plot in Excel (see Appendix 2).
30	Each data set must have its own distinguishable symbols or lines. Choose a dark color for both the line and the symbol.
31	Use scientific notation when numbers are very large or very small.
32	Units are needed.
33	Number format is incorrect. Do not start sentences with a number (write out the number). Make sure decimal numbers less than one start with zero, e.g. 0.1 mL, *not* .1 mL.

34	Discussion section requires an explanation of the results.
35	Discussion section requires a comparison of your data to what is known from the literature. Science is not done in a vacuum!
36	Wrong reference format. See Chapter 4 for specific examples.
37	All citations in the text must be listed in the References section.
38	All references must be cited in the text.

Laboratory Report Mistakes

Chapter 7

Poster Presentations

Posters are a means of communicating research results quickly. They provide a great opportunity to get feedback about preliminary data and ideas. **Poster sessions** are often held at large national meetings, and they allow you to meet other scientists in an informal setting.

Why Posters?

Scientists who attend poster sessions constitute a much larger audience than the one attracted to a journal article on a particular topic. Thus, your goal is to produce a poster that not only attracts experts in your subdiscipline, but also the much larger group of scientists with tangential research interests. The latter group provides a unique opportunity for you to learn about applications of your work to other research areas (and vice versa), spurs scientific creativity, and prompts you to apply an interdisciplinary approach to problem-solving.

Posters are *not* papers; they rely more on graphic illustrations than on text to present the message. It is not necessary to supply as many supporting details as you would for a paper, because you (the author) will be present to discuss details one-on-one with interested individuals. Too much material may even discourage individuals from reading your poster.

An appropriate poster presentation should fulfill two objectives. First, it must be esthetically pleasing to attract viewers in the first place. Second, it must communicate the methods, results, and conclusions clearly and concisely.

Poster Format

Size

Poster boards come in many sizes. Check with the conference organizer regarding minimum and maximum sizes. For poster sessions in your class, ask your instructor about appropriate materials and sizes.

Font (Type Size and Appearance)

Remember that most readers of your poster will be 3 to 6 feet away, so the print must be large and legible. Sans serif fonts like Arial are good for titles, but serif fonts like Times and Palatino are much easier to read in extended blocks of text. The serifs (small strokes that embellish the character at the top and bottom) create a strong horizontal emphasis, which helps the eye scan lines of text more easily.

Make the title Mixed Type or ALL CAPS in 72 point **bold**. Mixed type has the advantages of being easier to read and taking up less space than all caps. Do not use all caps if there are case-sensitive words in the title, such as pH, cDNA, or mRNA. Limit title lines to 65 characters or less.

Times
ARIAL

Authors' names and affiliations should be 48 or 36 point **bold**, serif font, mixed type:

Times 48 Pt

The section headings can be 28 point **bold**, serif font, mixed type:

Times 28 Pt

The text itself should be no smaller than 24 point, serif font, mixed type, and *not* in bold:

Times 24 Pt

Poster Esthetics

The success of a poster presentation depends on its ability to attract people from across the room. Interesting graphics and colored photos are good attention-getters, but avoid "cute" gimmicks. Present your poster in a serious and professional manner so people will take your conclusions seriously.

Organize the layout so that information flows from top left to bottom right. Align text on the left rather than centering it. The smooth left edge provides the reader with a strong visual guide through the material.

Avoid crowding. Large blocks of text turn off viewers; instead, use bullets to present your objectives and conclusions clearly and concisely. Use blank space to separate sections and to organize your poster for optimal flow from one section to the next.

Use appropriate graphics that communicate your data clearly. Use three-dimensional graphs *only* for three-dimensional data.

Colored borders around graphics and text enhance contrast, but keep framing to a minimum. Framing is the technique whereby the printed material is mounted on a piece of colored paper, which is mounted on a piece of different-colored paper to produce colorful borders. Use borders judiciously so that they do not distract from the poster content.

Nuts and Bolts

To affix text and figures to the poster board, use adhesive spray or glue stick. These tend to have fewer globs and bulges than liquid glues.

Ask the conference organizer (or your instructor) about how posters will be displayed at the poster session. Some possibilities include a pinch clamp on a pole, an easel, a table for self-standing posters, and cork bulletin boards to which posters are affixed with pushpins.

Poster Content

Posters presented at large national meetings should be organized so that readers can stand 10 feet away from the poster and get the take-home message in 30 seconds or less. Because of the large number of sessions (lectures) and an even larger number of posters, conference participants often experience "sensory overload." Thus, if you want your poster to stand out, make the section headings descriptive, the content brief and to the point, and the conclusions assertive and clear.

Posters for a student audience in the context of an in-house presentation should follow the same principal of brevity, but may retain the sections traditionally found in scientific papers. These include

- Title banner
- Abstract (optional—if present, it is a summary of the work presented in the poster)
- Introduction
- Materials and Methods
- Results
- Discussion or Conclusions
- References (these should be brief or omitted on the poster and provided on a handout)

Title Banner

Use a short, yet descriptive, title. This is the first and most important section for attracting viewers, so try to incorporate your most important conclusion in the title. For example, **Gibberellic Acid Makes Dwarf *B. rapa* Grow Taller** is more effective than **Effect of Gibberellic Acid on Dwarf *B. rapa***.

The title banner should be at the top of the poster and in 72 point bold font, mixed type, or all caps, 65 characters or less on a line. Underneath the title, include the authors' names, in alphabetical order, and the institutional affiliation(s). Use 48 point bold, mixed type for the authors' names.

Introduction

Instead of this conventional heading, consider using a short statement of the topic or introduce the topic as a question. Under the heading, briefly explain the existing state of knowledge of the topic, why you undertook the study, and what specifically you intended to demonstrate. A bulleted list of objectives may be a good way to present some of this information.

Materials and Methods

Present the methods you used to investigate the problem in enough detail so that someone competent in basic laboratory techniques could repeat your experiments. You might write the basic approach as a series of bulleted statements, and then provide more details in the subsequent text. Be both *brief* and *thorough*.

Results

The Results section of a poster consists mostly of graphic illustrations (pictures, tables, and graphs) and a minimum of text. Poster viewers do not have time to read the results leisurely, as they do with a journal article. An effective presentation of the results should announce each result with a heading, provide a graphic illustration that supports the result, and use text sparingly as a supplement to the graphics.

Figures are a summary of the raw data and are constructed so that viewers can appreciate both the general patterns of the data and the degree of variability that they possess. Written text should concentrate on general patterns, trends, and differences in the results, and not on the numbers themselves. For example, the reader can visualize "the concentration of chlorophyll increased initially, and then leveled off after 10 days" much more easily than "the chlorophyll concentration was 9×10^{-5} units on the fourth day, increased to 6×10^{-4} units on the tenth day, and then stayed about the same at 4×10^{-4} units on day 21."

Avoid using tables with large amounts of data; if you think the data are important, prepare the table to give as a handout. Your job is to sort through the data and come up with the take-home message. Do not use flip-out charts, in which one table or figure is displayed beneath another.

Figures may contain some statistical information including means and standard error and minimum and maximum values, where appropriate. Make the data points prominent and use a simple vertical line without crossbars for the error bars. Use the same-sized font for the axis legends and the key as you use for the text (24 pt). Do not present the same data in both a table and a figure.

Graphic illustrations in posters do not need a caption (number and title). Instead, the graphs and pictures are integrated so that they immediately follow the text in which they are first described.

Edit the text ruthlessly to remove nonessential information about the graphics. A sentence like "The effect of gibberellic acid on *B. rapa* is shown in the following figure." is nothing but deadwood. On the other hand, "*B. rapa* plants treated with gibberellic acid grew taller than those receiving only water" informs the reader of the result.

Remember to leave room for blank space on a poster. Space can be used to separate sections and gives the eyes a rest.

Discussion or Conclusions

Interpret the data in relation to the original objective or hypothesis and relate these interpretations to the present state of knowledge presented in the Introduction. Discuss any surprising results. Discuss the future needs or direction of the research. Where appropriate, identify sources of error and basic inadequacies of the technique. Do not cover up mistakes; instead, suggest ways to improve the experiment if you were to do it again. You may also speculate on the broader meaning of your conclusions in this section. Use bullets to help you present this information concisely.

Literature Citations

In scientific papers, it is common to cite the work of others, particularly in the Introduction and Discussion sections. The full references are then given in the References section at the end of the paper. Posters, on the other hand, are informal presentations that do not need to contain all the supporting details. Scientists who visit your poster are likely to work in the same field and probably are already familiar with most of the literature, so a long list of references would only waste valuable space on your poster. Even visitors who have a general knowledge of your topic but who work in a different subdiscipline are not interested in the details. They are interested mainly in how your approach or findings might help them improve their methodology or provide insight into their work. (Discussions with scientists in this second group are valuable to you because they provide a different perspective and may help you see applications to other subdisciplines.)

Student presentations in an in-house setting are different from poster sessions at large national meetings, because students usually do not have the background or the familiarity with the literature that career researchers have. Researching your topic is part of the scientific method, however, and presenting your work in the context of the published literature is part of good science. Thus, your instructor may ask you to include literature citations on your poster or to provide a list of references on a handout. See Chapter 4, "References," for proper citation and reference format.

Presenting Your Poster

Authors should prepare a 5-minute talk explaining their poster; anticipate questions from students and instructors, and prepare appropriate answers. After this presentation, at least one author must be present at his or her poster at all times to answer questions from the session participants. The more you interact with your audience, the more feedback you are likely to receive on your work. (On the other hand, do not throw yourself on passersby who demonstrate little interest.)

WORD PROCESSING BASICS IN MICROSOFT WORD

These instructions were written specifically for Word 97. However, when it comes to carrying out the tasks described in this appendix, there are not many differences between Word 97 and Word 2000. The most notable differences have to do with the menu bar and toolbar displays (see Section 2.1). Screen displays are from Word 97 except where noted.

Path notation for reaching a particular command or option is shown relative to a command on the menu bar, toolbar, or task bar. For example, File I Page setup I Paper size I Letter means "Open the File command (on the menu bar), then click Page setup, then select Paper size, and choose Letter."

1. Good Housekeeping Habits

1.1 File Organization

You can expect to type at least one major paper and perhaps several minor writing assignments in every college course each semester. This amounts to a fair number of documents on your hard disk, zip disk, or floppy disk. Good file organization is important for several reasons:

- When you are writing papers in different subjects at different times during the semester, you need to have a simple, logical system that will allow you to find the files you are working on. Nothing is more frustrating than to have a deadline and waste precious time searching for the wanted file.

- If your instructor informs you that he/she never received your paper, you will be expected to print out another copy. Excuses such as "I can't find the file" or "I accidentally erased it" are not viewed favorably.

- If you are asked to be a teaching assistant for an introductory course, it is helpful to be able to refer to your old lab reports when students ask you questions. The compactness and convenience of an electronic copy can't be beat.

When setting up a filing system on your computer, think of the computer as a large filing cabinet, with the drawers representing the computer's drives. Typically the hard drive is the C: drive, but newer computers may have the hard drive partitioned into several drives, designated C:, D:, E:, and so on. The floppy drive is usually the A: drive, and, if present, the CD (compact disk) and zip drives are given other letter designations.

In a paper filing cabinet, the drawers contain folders, the folders may be further subdivided into subfolders, and the subfolders contain the actual documents. Word creates a "My Documents" folder on the hard drive as a possible location for storing your files. You can create subfolders within "My Documents," or you can create folders in addition to "My Documents."

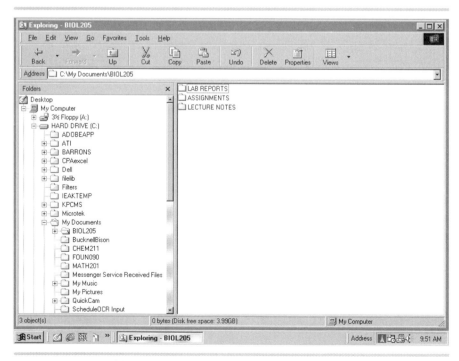

Figure A1.1 Possible way to organize your files using folders and subfolders

One way to organize your files is by course number (Figure A1.1). Within each course, you could create subfolders for different assignments, such as lecture notes, homework, or lab reports. If you expect to have only a few assignments that require work on the computer, you may not even need to make subfolders.

There are two easy ways to create folders or subfolders. One way is to use Windows Explorer, and the other way is to make folders within Word.

Creating folders in Windows Explorer. For PCs, click the Start button on the taskbar in the lower left corner of the screen. Then select Programs | Windows Explorer. Use the arrows to scroll up or down to the drive where you want to create a new folder (see Figure A1.1). Double-click the drive to display the existing folders. If you want to add a new folder on this level, select File | New Folder from the menu bar. Give the folder a unique, meaningful name that you will be able to recognize even years from now. To add a subfolder to an existing folder, first double-click the folder, and then select File | New Folder from the menu bar.

Creating folders in Word. You must have a Word document open to use this option. To open a Word document, open Word by clicking the Word icon on your screen. Alternatively, find Microsoft Word in Windows Explorer and double-click.

When you open Word, a blank document appears. At the top of the screen, there is a Title bar and a Menu bar and, if your Word program has been set up to display them, two Toolbars (Figure A1.2).

To create a new folder in Word:

1. Click File | Save As on the menu bar. The Save As dialog box appears.

2. Click the down arrow next to the Save in text box to locate the folder in which you want to create a subfolder.

3. Click the Create New Folder icon. The New Folder dialog box appears (Figure A1.3).

Figure A1.2 Screen display of a blank Word document

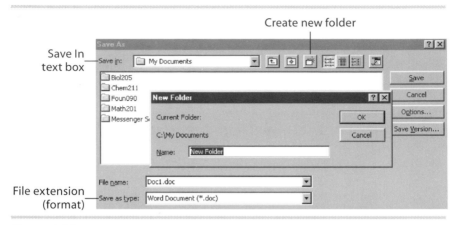

Figure A1.3 Save As dialog box allows you to create folders and subfolders in Word

> **4.** The path (directory) you selected is shown above the text box where you can enter the name of the new folder. If this is not the correct path, click Cancel.
>
> **5.** If the path is correct, type a name for the new folder and click OK.

The new folder now appears in the Save As dialog box.

1.2 Saving your Files

One of the most important tasks you must learn when working on the computer is saving your files. When you write the first draft of a paper by hand, you have tangible evidence that you have done the work. When you type something on the computer, however, your work is unprotected until you save it. That means that if the power goes off or the computer crashes before you save the file, you have to start again from scratch. Hopefully it won't take the loss of a night's work to convince you to make sure you know how to save your files!

Save your work often. The more difficult or complicated the text, the more often you should save it. Think about how long you would need to retype the text if it were lost and if you can afford to spend that much time redoing it.

Saving files manually. Although AutoSave provides some insurance against lost work, saving your file manually produces a permanent file that will still be there if the power goes off or the computer crashes.

To save your file for the first time:

1. Click File | Save as on the menu bar (see Figure A1.2). The Save As dialog box appears (see Figure A1.3). Do not wait until you are finished typing the document to save the file. Save it after you have written the first sentence. Even though Word and other word processors save the file in a temporary memory, it is not protected unless it has been saved in a permanent form.

2. You have to make three important decisions when saving a file for the first time:
   - Location (drive, folder, and subfolder, if applicable)
   - File name
   - File extension (format)

The default location where the current document will be saved is shown in the Save In text box. To change to another folder or another drive, click the arrow next to the text box. Double-click the drive, folder, or subfolder until you arrive at the location where you want to save the current document. Only the folders for the selected drive will be displayed. To see the folders in another drive, double-click the desired drive.

You will need to choose a file name. The name suggested by Word is shown in the File Name text box near the bottom of the dialog box. If necessary, replace Word's suggestion with a unique, meaningful name. Use as few words (characters) as possible, because long file names may be truncated to eight characters in earlier versions of Windows (e.g., ProteinAssayLab.doc becomes Protein~.doc).

Sometimes it is desirable to keep track of the different versions of a particular document. In that case, amend the file name with numbers or letters (e.g., Protein1.doc, Protein2.doc, etc.).

Windows machines automatically assign a file extension (the three or four letters following the period in the file name) based on the program you are working in. It is best not to change the extension unless you expect to open the file in a different program. Some common file extensions are shown below. When you have entered the location, file name, and file extension, click Save to save the file and close the dialog box.

Program	Extension
Word	.doc
Excel	.xls
PowerPoint	.ppt
Adobe Acrobat	.pdf
Images	.jpg, .tif, .gif
HTML	.html, .htm, .htx

To save the same file subsequently, click the Save command (floppy disk icon) on the Standard toolbar, or File | Save on the menu bar.

To save the same file at a different location, under a different name, or with a different file extension

1. Click File | Save as on the menu bar or press F12. The Save As dialog box appears (see Figure A1.3).

2. Change the location, name, or extension as needed.

3. Click Save to save the file and close the dialog box.

AutoSave. AutoSave automatically saves the file you are currently working in at specified intervals. It should be considered a backup to (not a substitute for) saving the file manually.

1. To activate AutoSave, click Tools | Options on the menu bar (see Figure A1.2).

2. Select the Save tab in the Options dialog box (Figure A1.4).

   Select
   - Allow fast saves
   - Allow background saves

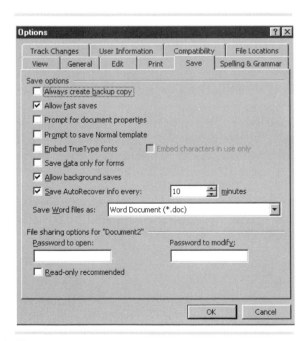

Figure A1.4 Options dialog box, Save tab

- Save AutoRecover info every: ___ Minutes, and use the arrow to specify the interval (e.g., 10 minutes).
- Save Word files as: Word Document (.doc)

The Allow fast saves option makes saving files faster by recording only those changes made in the document during the current session. The changes are then added (electronically) to the end of the file, instead of incorporating them into the file at the location where they were made. Fast saves increase the size of the document, so it is a good idea to turn off this option when you save the final version of a file. Disallowing fast saves forces Word to do a full save, in which the changes are incorporated properly, and the file size is reduced.

AutoSave saves the document in a recovery file. This feature is not a substitute for manually saving documents, because the recovery file is deleted when you save or close a document. If the power goes off, the recovery file still exists. When you restart Word, Word opens the recovery file so that you can save it manually. If you do not save the recovery file, it is deleted.

Backup files. It is very, very important to make at least one backup copy of your file while you are writing an important paper. The backup file should be saved to a disk, not on your hard drive. If your hard drive goes down, if you have a backup copy on a zip disk or floppy disk, you have not lost your work and can keep working on your paper on another computer.

To make a backup file, follow the instructions in Section 1.2. Change the Location to either the floppy disk drive or the zip disk drive, depending on your computer's setup.

1.3 Protecting your Files Against Viruses and Worms

Keep your antivirus software up to date by downloading the most recent software from the company's Internet site. New viruses and worms infect someone's computer every day, so you must aggressively protect your system and all the time invested in writing your papers. An ounce of prevention is worth a pound of cure, as the saying goes.

1.4 Removing Files

If you use your computer a lot, and especially if you have large files and graphics, the disk space fills up fast. Your computer will give you a message when you are running out of disk space. That is a sign that you need to remove files from the hard drive(s).

If you have already backed up an inactive file (one on which you are no longer working) on floppy disk or zip disk, then you can delete this file from the hard drive. Also check for files in the Recycling Bin, for temporary files and files attached to e-mail messages, and images saved from the Internet. Copying them to floppy or zip disk and then deleting them from the hard disk frees up valuable space on your computer.

If you need to remove a lot of files from your computer, use Windows Explorer:

1. For PCs, click the Start button on the taskbar in the lower left corner of the screen.
2. Select Programs | Windows Explorer.
3. Start with the C: drive, and look in each folder for inactive files.
4. If you want to save a file for future reference, copy it to a floppy or zip disk as follows:
   a. Insert a floppy disk or zip disk into the respective drive.
   b. Click Start | Programs | Windows Explorer to open a second Explorer window.
   c. Use the arrows to scroll up or down to the floppy or zip drive.
   d. Double-click the drive to display the contents of the disk.
   e. Click the file on the hard drive that you want to copy to the floppy or zip disk.
   f. Click Edit | Copy from the menu bar. Alternatively, right-click the mouse and select Copy.
   g. Click the Explorer window displaying the contents of the floppy or zip disk.
   h. Click Edit | Paste from the menu bar. Alternatively, right-click the mouse and select Paste. The copied file will be displayed in the window for the floppy or zip drive.
   i. An even faster way to copy is to click the file to be copied and drag it to the new location. This method replaces steps f through h.
   j. Return to the hard drive, to the file you just copied. Click the file, and then File | Delete on the menu bar. A message asks you to confirm that you really want to delete the file. Click "Yes," and the file will disappear from the window.
5. If you want to copy several consecutive files, click the first file, hold down the Shift key, click the last file, and repeat steps 4f through 4j.
6. To delete files without copying them, simply click the unwanted file, click File | Delete and "Yes" to confirm that you want to delete it.

You can also delete individual files in Word as follows:

1. Open Word.
2. Click File | Open on the menu bar.
3. Locate the file you want to delete (you may have to change drives or folders).
4. Right-click the unwanted file, and select Delete.

2. Formatting a Document

All scientific journals have a section devoted to instructions for authors. These instructions tell the author exactly how to format a paper, so that all the papers in the journal have a uniform appearance. If your instructor has not given you specific instructions, the layout in Table A1.1 will lend a professional look to your lab report or scientific paper. If you are already familiar with Word, the Quick Reference tells you where to make these formatting adjustments. Detailed instructions are given in the sections following the table.

2.1 Menu Bar and Toolbars

When you open Word, a blank document appears. At the top of the screen, there is a Title bar and a Menu bar and, if your Word program has been set up to display them, two Toolbars. The screen display in Word 97 looks like Figure A1.5a.

If the Standard and Formatting toolbars are not displayed, go to the menu bar and select View | Toolbars. Click the check boxes next to Standard and Formatting, then click Close. The View | Toolbars command shows the most commonly used toolbars (as decided by Word). To view *all* of Word's default toolbars, click Tools | Customize | Toolbars.

Toolbars contain icons (pictures) that are shortcuts to menu commands. If you place the cursor (arrow) over a particular icon (do not

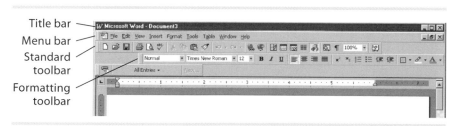

Figure A1.5a Screen display of a blank Word 97 document

TABLE A1.1 Giving your document a professional appearance

FEATURE	LAYOUT	QUICK REFERENCE
Paper	8½ x 11" (or DIN A4) white bond, one side only	File | Page setup | Paper size | Letter
Margins	1½" left and right; 1" top and bottom	File | Page setup | Margins
Font size	12 pt (points to the inch)	Formatting toolbar | Font size
Typeface	Times Roman (serif) or Helvetica (sans serif). Serif fonts are easier to read for most people, but sans serif fonts are favored in graphics.	Formatting toolbar | Font See also Format | Style
Pagination	Arabic number, top right on each page except the first	Insert | Page numbers
Page Break	To maintain section continuity	Ctrl+Enter or Insert | Break | Page Break
Spacing	Double, except title, list of authors, and figure and table titles (which should be single spaced)	Format | Paragraph (Indents and Spacing tab) | Line spacing
New paragraph	Indent 0.5"	Format | Tabs | Default
Aligning text	**Justification** is Left, ragged right, or Justify/even edges	Format | Align left or Justify
	Indent entire paragraph to set off from rest of text	Format | Paragraph | Left and right indentation
	Tabs	Format | Tabs
	Table: Create a table	Table | Insert table
	Lists that begin with **Bullets** or **Numbers**	Format | Bullets and Numbering
Title page (optional)	Title, authors (your name first, lab partner second), class, and date	
Headings	Align headings for Abstract, Introduction, Materials and Methods, Results, Discussion, and References on left margin or center them. Use consistent format for capitalization.	Format | Style | Heading 1.
Subheadings	Use sparingly and maintain consistent format.	Format | Style | Heading 2, Heading 3, etc.
References	Citation-Sequence System: Make a numbered list.	Format | Bullets & Numbering | Numbered tab
	Name-Year System: Use hanging indent to list references.	Format | Paragraph (Indents and Spacing tab) | Special | Hanging by 0.25"
	Both systems: Use accepted punctuation and format.	
Assembly	Pages in order, staple top left corner	

click any mouse buttons) on either toolbar, a description of that icon will appear. For example, if you place the cursor on the fourth icon from the left on the Standard toolbar (printer icon), the Print command, with the

Word 2000 Update

In the Word 2000 screen display, the Standard and Formatting toolbars are combined into one long row (Figure A1.5b).

Although the single row of toolbar buttons provides a little extra space in the document window, not all of the buttons are displayed. If you are accustomed to seeing all of the buttons, you may find it a nuisance to have to click More Buttons (down arrow on the right side of the toolbar) every time you want to see the hidden buttons.

To display the Standard and Formatting toolbars on two rows, select Tools | Customize | Options, and deselect the *Standard and Formatting Toolbars share one row* check box.

Word 2000 has also changed the pulldown menus for the menu bar commands. Instead of listing all of the commands in the pulldown menus (as in Word 97), the so-called adaptive menus show only the basic commands (as decided by Word), and then add the commands you use in the current session.

If you are accustomed to having all of the commands available for each menu command, you will find the adaptive menus a nuisance. If you are using Word for the first time, it may be difficult for you to find certain commands; either you have to wait five seconds before Word displays the rest of the commands, or you have to click the down arrow to see them.

To turn off the adaptive menus, click Tools | Customize | Options, and deselect the *Menus show recently used commands first* check box.

Figure A1.5b Screen display of a blank Word 2000 document

default printer in parentheses, will appear. The Print command is also found on the menu bar under File | Print.

Note: Shortcuts offer only one option from the corresponding menu command. For example, if you click the printer icon on the standard toolbar, the entire document will be printed on the default printer. If you want to print single pages or use a different printer, you must go to the menu bar and select the desired options from the File | Print drop-down menu.

The menu bar commands that are important for formatting your document are File, Insert, and Format (Figure A1.6).

2.2 Paper Size

Click File | Page Setup on the menu bar. You will find four tabs: Margins, Paper Size, Paper Source, and Layout (Figure A1.7). Select Paper Size. Letter (8.5 x 11") is the standard paper size in the United States; most other countries use A4 (8.27 x 11.69"). Select Letter. Then select Portrait for the orientation (so that the paper is vertical, rather than horizontal).

2.3 Margins

Click File | Page Setup on the menu bar (see Figure A1.7). Now click the Margins folder. Set the top and bottom margins to 1" and the left and right margins to 1.5". These generous left and right margins will give your instructor room to make comments. The header and footer are generally set to 0.5" from the edge of the paper. This is where page numbers are printed.

2.4 Font Style and Point Size

Click Format | Font on the menu bar. Under Font, select a conventional font such as Times Roman or Arial (serif fonts such as Times Roman are easier to read for most people than sans serif fonts such as Arial). Under Font Style, select Regular (i.e., not bold, italics, or underline), and under Size select 12 pt.

Figure A1.6 Menu bar with arrows on commands used to format a Word document

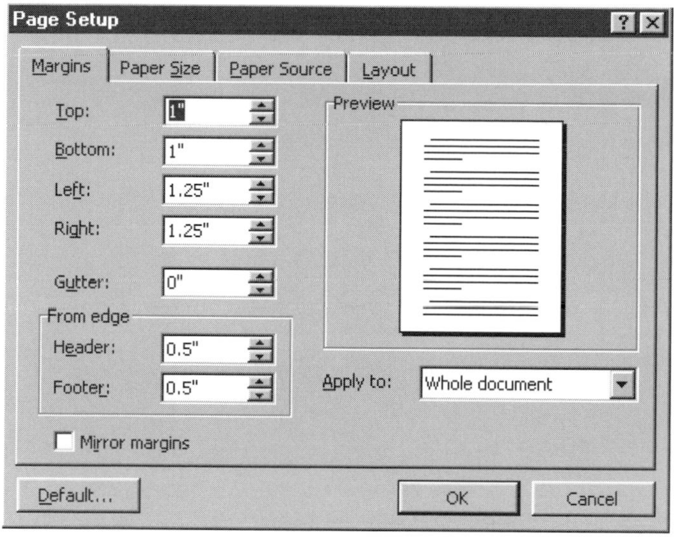

Figure A1.7 Page Setup dialog box

These formatting options are also available on the Formatting toolbar (Figure A1.8).

2.5 Page Numbers

Click Insert | Page Numbers on the menu bar to open the dialog box (Figure A1.9).

In the Page Numbers dialog box, select the position (top or bottom of the page) and the alignment (left, center, right, inside, or outside) of the page numbers. Journal articles tend to have the page numbers placed at the bottom, on the outside margins. For your course work, however, top right is fine. Be sure to de-select Show number on first page, as it is not customary to put a number on the first page. All pages of your paper except the first page should have a page number, so that you can easily determine the order of the pages and whether all pages are present.

Figure A1.8 Formatting toolbar

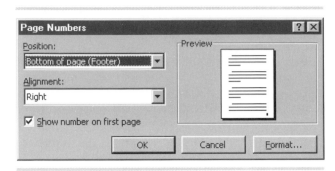

Figure A1.9 Page Numbers dialog box

By convention, Arabic numerals are used to number the pages. This style is the default page number format in Word. In some special situations (e.g., Table of Contents or Preface of a book), Roman numerals or letters are used. These alternative formats can be selected by clicking the Format button.

The font of the page number should be the same as the font of your text. Because page numbers are part of the header or footer, you can check the page number font by displaying the header and footer as follows:

- Click View | Header and Footer, or
- Click View | Page Layout (called Print Layout in Word 2000)

Double-click the page number to view the current font, font style, or font size. Modify the format by using the corresponding options on the Formatting toolbar (see Figure A1.8).

To modify the page number position or alignment at any time, click Insert | Page Numbers on the menu bar.

2.6 Page Breaks

A **page break** starts the text following the break on the next page. For example, let's say that you start a new section of your lab report or scientific paper near the bottom of a page. Perhaps the section heading fits, but the rest of the text gets thrown onto the next page. To make sure that the section heading is not separated from the rest of the text, position the cursor just in front of the heading. Click Insert | Break | Page break. A faster way to do this is to put the cursor just in front of the heading and press Ctrl+Enter on the keyboard.

2.7 Line Spacing

Line spacing refers to the amount of space between lines on a page and is found under Format | Paragraph (Indents and Spacing tab) on the menu bar. To double-space your paper, select Double under Line spacing in the Paragraph dialog box (see Figure A1.10). If some parts of your paper are to be single-spaced, highlight these sections after you have typed them, then change spacing to Single using the same procedure. To highlight text, hold down the left mouse button, and drag it across the text from the first character to the last.

2.8 Aligning Text

There are a number of different ways to align text in a Word document. The ones you are most likely to use are as follows:

- Justification (right, left, center, or full). **See Section 2.9.**
- Indenting an entire paragraph so that it is set off from the rest of the text. **See Section 2.10.**
- Tabs. **See Section 2.11.**
- Table columns (creating a table). **See Section 2.12.**
- Lists that begin with numbers or bullets. **See Section 2.13.**

Never use the space bar to align text, because variation in character size will lead to variation in vertical text alignment as well.

2.9 Justification

Justification is the adjustment of the lines so that they are aligned on the left margin (with ragged right edge), centered between the margins, aligned on the right margin (with ragged left edge), or aligned both on the left and right margin (with even edges). Under Format | Paragraph, in the Paragraph (Indents and Spacing tab) dialog box, select either Left or Justify next to Alignment (Figure A1.10). These options are also available on the Formatting toolbar (see Figure A1.8).

2.10 Indentation

Indenting a block of copy allows you to change the margins, so that only that block of copy will be set off from the others.

To indent copy, click Format | Paragraph on the menu bar. In the Paragraph dialog box (see Figure A1.10), set the left and right Indentation as

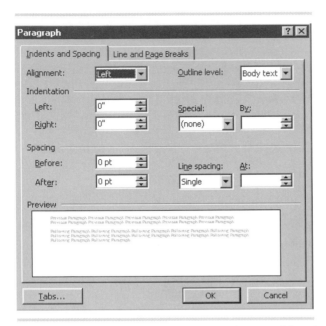

Figure A1.10 Paragraph dialog box, Indents and Spacing tab

desired (for example, 1"). When you type the text, it will be set off from the rest of the text with wider margins. Pressing Enter returns the indentation to the default values.

If you have already typed a passage that you later decide you want to indent:

1. Highlight the passage. To highlight text, hold down the left mouse button, and drag it across the text from the first character to the last.
2. Click Format | Paragraph.
3. Set the desired indentation in the Paragraph dialog box.
4. Click OK.

2.11 Tabs

Tab stops allow you to jump a set number of spaces forward on a line. It is always preferable to use tab stops rather than the space bar to align text in columns. Because character spacing varies with the size of the characters in a font, different words, even though they have the same number of characters, occupy different spaces on a line. Adding more spaces will not change this mismatch. Instead, put the words on a tab stop, which defines the vertical alignment regardless of the characters. Remember, **never use the space bar to align text in columns!**

To adjust tab stops, select Format | Tabs on the menu bar. The default tab stops are given in the top right corner of the Tabs dialog box and will most likely be 0.5". Change the default if desired, or add individual tab stops as needed.

The indentation that designates a new paragraph is determined by the first tab stop in the default setting. The customary paragraph indentation is 0.5".

2.12 Creating a Table

If you want to align your text in multiple columns, make a table. Creating a table is easy, and navigation among the cells (table entries) is simple. You don't have to worry about column width, because it is adjusted automatically depending on the number of columns. If you want to change the width of any of the columns later, that is also easy to do.

To create a table:

1. Position the cursor where you want to insert the table.
2. Click Table | Insert Table on the menu bar. Enter the number of columns and rows.
3. To apply a particular format to the table, click the AutoFormat button. By convention, tables in scientific papers do not have vertical lines to separate the columns (other than perhaps the first column from the rest). Furthermore, most scientific journals publish articles in black and white, so avoid formats that use color. Simple 1, Simple 2, or Classic 1 would be appropriate table formats for your scientific paper.
4. Click OK in the Insert Table dialog box.

A blank table appears with the cursor in the first cell.

Tables with no borders. To align text in columns without showing any lines:

1. Select all the cells in the table.
2. Click Format | Borders and Shading on the menu bar.
3. On the Borders tab under Setting, click None.
4. Click OK.

Navigation in tables. To **jump** from one cell to an adjacent one, use the arrow keys. To move forward across the row, use the Tab key. Note: If you press Tab when you are in the last cell of the table, Word adds another row to the table.

To insert a tab stop in a table cell, press Ctrl+Tab.

Gridlines and Borders. If you did not select a format from the Auto-Format list, Word puts a single, thin line around each cell in the table (so that the table looks like a grid). To remove the vertical lines (as per scientific journal convention):

1. Select all the cells in the table.
2. Click Format | Borders and Shading on the menu bar.
3. In the Preview box, click each vertical line inside the table to remove it.
4. Click OK. The vertical lines will still be visible, but they will not be printed.

Inserting and Deleting Rows or Columns. To insert a new column:

1. Position the cursor in the column to the right of where you want to insert the new column.
2. Click Table | Insert Columns on the menu bar.

To insert a new row:

1. Position the cursor in the row below where you want to insert the new row.
2. Click Table | Insert Rows on the menu bar.

To delete a column or row:

1. Position the cursor in the column or row you want to delete.
2. Click Table | Delete Cells on the menu bar.
3. Select Delete entire row or Delete entire column in the Delete Cells dialog box.
4. Click OK. The row or column will be deleted.

Changing Column Width or Row Height. To change the width of a column, position the cursor on one of the vertical lines so that ⫝⃒ appears. Then hold down the left mouse button and move the line to make the column the desired width.

To change the height of a row, position the cursor on one of the horizontal lines so that ⫩ appears. Then hold down the left mouse button and move the line to make the row the desired height.

Other Table Commands. Position the cursor in the table, and then click Table on the menu bar. Other commands for customizing your table are shown in the pulldown menu (Figure A1.11). The Table menu in Word 2000 has a few additional commands.

You can format the text in each cell just as you would format text outside the table. For example, to center the column headings, highlight the top row and select the Center command from the Formatting toolbar. Alternatively, click Format | Paragraph, Alignment: Center in the Paragraph dialog box (see Figure A1.10). To print the headings in bold-face, highlight the headings and select the Bold command from the Formatting toolbar. To align numbers, highlight the entire column and click the Align right command on the Formatting toolbar. (To highlight text, hold down the left mouse button, and drag it across the text from the first character to the last.)

2.13 Lists That Begin with Numbers or Bullets

A **numbered list** may be used when you want to list references in the Citation-Sequence system or steps in a procedure or organize your thoughts first as an outline. A **bulleted list** is handy for highlighting or summarizing the main points of a paper or poster, when chronological order is not important.

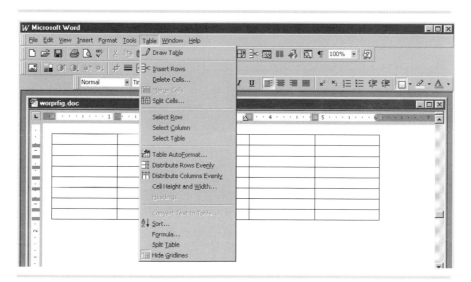

Figure A1.11 Table commands

Even though it is practical to number the steps in a procedure for your own reference, this is never done in scientific papers. The procedures are always written in paragraph form in the Materials and Methods section, never as a numbered list.

Bulleted lists. To make a bulleted list:

1. Click Format | Bullets and Numbering on the menu bar. You will see three tabs in the Bullets and Numbering dialog box: Bulleted, Numbered, and Outline Numbered (Figure A1.12).

2. Choose the Bulleted tab to make an unnumbered list. You have a choice of eight bullets: none and symbols such as filled dots, open dots, arrows, and boxes. Click the one you want.

3. If you prefer a different bullet, click any box except none, and then press the Customize button at the bottom of the dialog box. In the Customize Bulleted List dialog box (Figure A1.13), click the Bullet button. This opens the Symbols dialog box, which allows you to change the font to any ASCII character.

4. Again in the Customize Bulleted List dialog box, click the Font button. This allows you to select the format of the text after the bullet. This is usually the same format as the rest of the document.

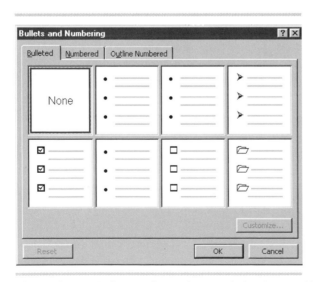

Figure A1.12 Bullets and Numbering dialog box, Bulleted tab

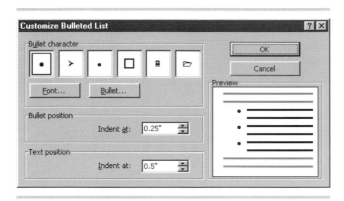

Figure A1.13 Customize Bulleted List dialog box

5. Finally, set the indentation of the bullet and the text. If you want the bullet to start on the left margin, set the indentation to 0". Set the text indentation 0.25" more than the bullet indentation. Click OK

6. When you have finished typing the list, press Enter, and then Backspace to delete the bullet on that line.

You can also type a list first, and then add bullets later on. To do this, highlight the list (To highlight text, hold down the left mouse button, and drag it across the text from the first character to the last.), and then select Format | Bullets and Numbering, and the desired format.

Numbered lists. If you want to make a numbered list:

1. Click Format | Bullets and Numbering on the menu bar. You will see three tabs in the Bullets and Numbering dialog box: Bulleted, Numbered, and Outline Numbered (see Figure A1.12).

2. Choose the Numbered tab (Figure A1.14). You have a choice of eight options: none and different styles of numbers or letters. Click the one you want.

3. If you prefer another option, click any box except none, and then press the Customize button at the bottom of the dialog box. In the Customize Numbered List dialog box, set the Number format, style, and position, and the Text position (0.25" more than the number position).

4. Back in the Bullets and Numbering dialog box (Numbered tab), select how you want the list numbered: Restart numbering, or Continue previous list. If you didn't have any previous numbered list of that format, these selections will not be available.

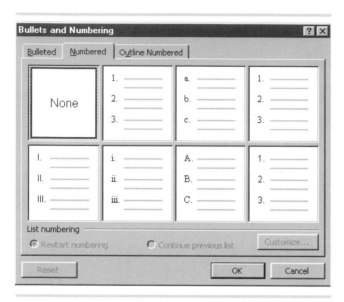

Figure A1.14 Bullets and Numbering dialog box, Numbered tab

2.14 Style

A **style** is a set of formats that can be applied to a few words, a paragraph, or a whole document. Look on the left side of the Formatting toolbar for a box containing the word *Normal* (see Figure A1.8).

Click the down arrow next to the word Normal to see what font typeface and size are currently specified for the normal text (running text of the document). In this pulldown menu, you can also see what styles are specified for different levels of section headings and for the header and footer. In order to change the style for the Normal text:

1. Click Format | Style (Figure A1.15).

2. Choose Normal from the choices on the left side of the Style dialog box.

3. The current format for the Normal style is described in the lower right area. For example, the Normal style used to type this text has the following description:

   > Times New Roman, 12 pt, English (United States), Character spacing 100%, Flush left, Line spacing single, Window/orphan control, Body text

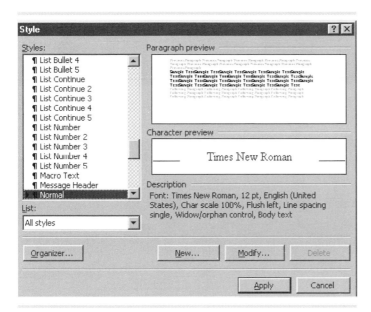

Figure A1.15 Style dialog box

If you want to change any of these formatting characteristics,
click the Modify button, and then Format at the bottom of the
Modify style dialog box. Change the desired features, and
press OK.

4. Back in the Style dialog box, click Apply, and the new features
   will be applied to all Normal text in your document.

To ensure that your section headings (Introduction, Materials and Methods, etc.) are uniform:

1. Click Format | Style (see Figure A1.15).

2. Select Heading 1 from the choices on the left side of the Style
   dialog box.

3. Click the Modify button to modify the characteristics of
   Heading 1 as needed (typeface should be the same as Normal,
   but font size could be larger, and the heading could be boldfaced to make it stand out).

4. Press OK.

5. Back in the Style dialog box, click Apply.

6. After you type a heading in your paper, highlight it, and then click the down arrow next to Normal to select Heading 1. The style you defined for Heading 1 is automatically applied to all the level 1 headings in your document.

Headings are also used to generate a Table of Contents, so you might find this feature useful for longer research papers.

To make sure the font of the page numbers is the same as the font for the body of the paper:

1. Click Format | Style.

2. Choose Header on the left side of the Style dialog box. (If the page number is in the footer, choose Footer.)

3. The current format for the Header style is described in the lower right of this dialog box.

4. If you want to change any of these formatting characteristics, select Modify, and then Format at the bottom of the Modify style dialog box. Change the desired features, and click OK.

5. Back in the Style dialog box, click Apply, and the new features will be applied to all Headers in the current document.

2.15 Templates

Templates specify the format of all Word documents. When you select File | New from the menu bar or the New Document icon from the Standard Toolbar, Word opens a new document based on the Blank Document template. This is the default template. To see the format properties specified in the default template,

1. Select File | New from the menu bar (Figure A1.16).

2. Double-click the Blank Document icon on the General tab.

3. Check the settings for each feature in Table A1.1.

4. Click Close.

If your instructor has specified a particular format for all your papers, you can customize an existing template for the format you need. Follow these steps:

1. Select File | New from the menu bar (see Figure A1.16).

2. Select the Template option under Create New.

3. Click the Blank Document icon on the General tab.

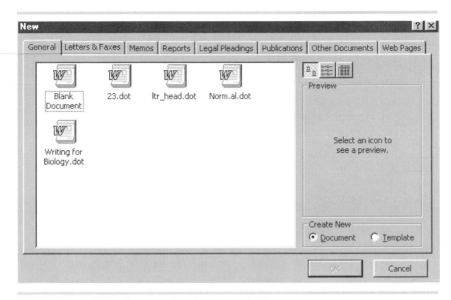

Figure A1.16 Templates under the general tab

4. Click OK.

5. Check the settings for each feature in Table A1.1, or for your instructor's specifications, and make changes where necessary.

6. Select File | Save As from the menu bar. The Save As type will automatically be Document Template (*.dot).

7. Type the name of the new template next to File name. For example, you might give the template the same name as your course.

Every time you type papers for that course, select File | New, and then click the course icon. Using the same template gives a consistent format to your documents.

3. Editing the Text

You will modify parts of your document while you are writing it, as well as afterwards during the editing and proofreading stages. Some of the most common modifications require the following tasks:

- Views
- Cut, copy, and paste
- Find and replace

- Spelling and grammar checker
- Word count
- Tracking changes made by peer reviewers

Use Table A1.2 if you are already familiar with these features. Detailed instructions follow the table.

3.1 Views

When you click View on the menu bar, you are given a number of choices that affect how your document appears on the screen (Figure A1.17). Normal is useful when you are interested in seeing only the text, but not the headers, footers, blank areas, or figures (images or graphs imported into the document). On the other hand, if you want a more realistic view of what the page will look like with headers, footers, blank areas, and figures (but no hidden characters), then use the Page Layout (called Print Layout in Word 2000) view. If you need to see the hidden characters, choose Outline or Master Document.

TABLE A1.2 Quick reference for editing commands

FEATURE	QUICK REFERENCE
Views	**Word 97:** View | Normal / Online / Page Layout / Outline / Master Document / Toolbars /Header and Footer / Zoom
	Word 2000: View | Normal / Web Layout / Print Layout / Outline / Toolbars / Header and Footer / Zoom
Zoom	Standard toolbar | Select percentage or another option
Cut, Copy, and Paste	Standard toolbar | Scissors icon, Two pages icon, Clipboard with page icon
Find, Replace	Edit | Find or Replace
Automatic spelling and grammar checker	Tools | Options | Spelling and Grammar
Manual, systematic spelling and grammar checker	Standard toolbar | $^{ABC}_{\checkmark}$ icon
Word count	Tools | Word Count
Tracking changes made by peer reviewers	Tools | Track Changes | Highlight Changes | Click Track changes while editing box

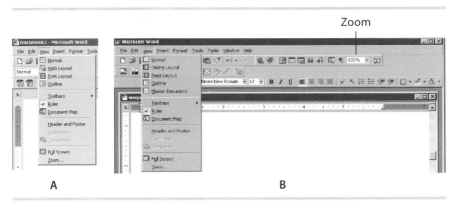

Figure A1.17 Comparison of View menus in Word 2000 (A) and Word 97 (B)

To see which toolbars are currently displayed, click Toolbars. Those with a check mark next to them are displayed.

To edit the header and footer, click Header and Footer.

To adjust document size on the screen, click Zoom. You can type in a percentage, or select an option from the list. Choose a size that makes it easy for you to read the characters on the screen. The Zoom box is also located on the Standard toolbar, at the top right of the screen (see Figure A1.17).

3.2 Cut, Copy, and Paste

To use the cut and copy functions, you must first highlight the text that you want to cut or copy. To highlight text, hold down the left mouse button, and drag it across the text from the first character to the last. Let go of the left mouse button, and click either the cut command (scissors on the Standard toolbar) or the copy command (two pages next to the scissors). Move the cursor to where you want to paste this text. It can be pasted into the same document or into another document. Click the paste command (clipboard with page next to the copy command).

3.3 Find and Replace

Sometimes you realize that you have been using the wrong word or wrong symbol throughout a paper. This error can be easily remedied using the Replace option. Click Edit | Replace on the menu bar. Type in the wrong word next to Find what, and the correction next to Replace with. If you are certain that the "wrong word" is wrong every time it appears in the document, then select "Replace all." If the "wrong word" may not be wrong every time, select "Find next." This allows you to preview the sentence in which the word was used and gives you the option *not* to replace it in every instance.

When you write the first draft of a paper, you may want to get all your ideas out first, without worrying about details or accuracy. It is important in subsequent drafts, however, to get the facts straight. You should devise your own way of marking the text that needs special attention. For example, you might type ?? after such text. Then when you want to go back and revise these sentences, you simply click Edit | Find, and type in ??.

3.4 Spelling and Grammar

There is absolutely no excuse for typos when Word offers you so many options for checking spelling and grammar. In versions of Word 97 and higher, there are three ways to correct spelling and grammar:

- AutoCorrect
- Automatic spelling and grammar checker
- Manual, systematic spelling and grammar check

AutoCorrect. The AutoCorrect function corrects common types of spelling mistakes as you type. To see the list of commonly misspelled words that Word corrects automatically, click Tools | AutoCorrect, and scroll through the words in the AutoCorrect tab. If you do not want Auto-Correct to change a particular keystroke combination (e.g., do not change (c) to ©), then select this combination on the list, and click the Delete button. You can also add words that you know you misspell frequently to the list.

If AutoCorrect changes text that you don't want changed while you are typing, click Edit | Undo AutoCorrect on the menu bar.

AutoCorrect may also be set to capitalize the first letter of sentences (a mistake if the word "pH" starts the sentence) and to display Internet and network paths with hyperlinks (Word 2000 only). To turn these functions off, click Tools | AutoCorrect, and deselect these check boxes.

Automatic spelling and grammar checker. The automatic spelling and grammar checker underlines in red the words that are not in Word's spelling dictionary, and underlines in green unusual sentence constructions.

To deal with a word underlined in red, click it with the *right* mouse button. A pop-up menu appears with commands and suggestions for replacements (Figure A1.18). Select "Add" after you have confirmed the correct

Insert a cuvette into the spectrophotometer.
curette
civet
covet
caveat
coveted
Ignore All
Add
AutoCorrect
Spelling...

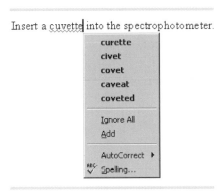

Figure A1.18 Dialog box for spelling suggestions and relevant commands

spelling in your textbook or laboratory manual. The red underline is deleted, and the word is ignored in the manual, systematic spelling and grammar check.

To deal with an unusual sentence construction underlined in green, click it with the *right* mouse button. A pop-up menu appears (Figure A1.19).

To tell Word how aggressively to check spelling and grammar as you type, click Tools | Options on the menu bar. In the Options dialog box (Figure A1.20), choose the Spelling and Grammar tab.

Check the boxes of the options you want. For example, under Spelling you might check:

- Check spelling as you type
- Always suggest corrections
- Ignore words in uppercase
- Ignore words with numbers
- Ignore Internet and file addresses

Under Grammar, you might select the following options:

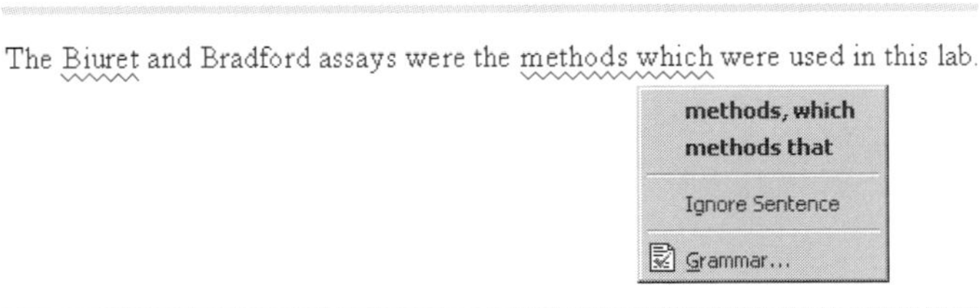

The Biuret and Bradford assays were the methods which were used in this lab.
methods, which
methods that
Ignore Sentence
Grammar...

Figure A1.19 Dialog box for editing unusual sentence constructions

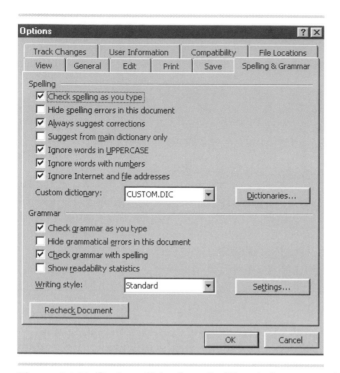

Figure A1.20 Options dialog box, Spelling & Grammar tab

- Check grammar as you type
- Check grammar with spelling

When you click the Settings box, you have the opportunity to select various grammar and style options.

Manual, systematic spelling and grammar check. To do a manual, systematic spelling and grammar check at any time, click Tools | Spelling and Grammar. The checker begins at the position of the cursor, and stops at every location at which it doesn't recognize a word or sentence construction. You then have the option of changing the word/construction, ignoring it, or adding the word to Word's dictionary. When the spell checker stops at a scientific word, consult your textbook for the correct spelling, as specialized terminology is not included in Word's dictionary.

As good as these features are, remember that a spell checker is no substitute for a human proofreader. Spell checkers cannot recognize mistakes of misuse, such as *there* and *their,* and they are not familiar with expressions such as *pH.* In fact, the spell checker may even try to get you to change a correctly used scientific word to a word that is in its repertoire, but that obviously makes no sense. Be vigilant and use common sense!

3.5 Word Count

The length of your lab report or paper depends on the detail expected by your instructor. Most instructors prefer not to specify a page length, stating simply that the work should be complete and concise. As a general rule, shorter is better.

If you do undergraduate research later on, you may be expected to submit a proposal of a certain length. If the specifications are given in terms of words, the word count feature is helpful. Click Tools | Word Count, and Word will tally the number of words in the document.

3.6 Tracking Changes Made By Peer Reviewers

It may not always be possible for you and your peer reviewer (classmate) to find a common time to go over your paper. E-mail makes the peer review process more convenient. You can send your paper to your peer reviewer in electronic format as an attached file, your peer reviewer can make comments directly in your file, and then the reviewer can send it back to you.

It is important for you to be able to distinguish your original text from the comments and changes suggested by your peer reviewer. After all, you are the author, who has the right to accept or decline the reviewer's suggestions.

In order to track changes made by your peer reviewer, open the document to be edited. Then click Tools | Track Changes and then Highlight Changes. In the Highlight Changes dialog box (Figure A1.21), check the Track changes while editing box. As long as this box is selected, anything someone types in the document will be underlined and be

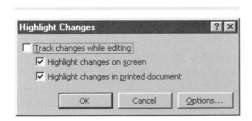

Figure A1.21 Highlight Changes dialog box

Word 2000 Update

Clicking the Options button in the Highlight Changes dialog box opens the following dialog box:

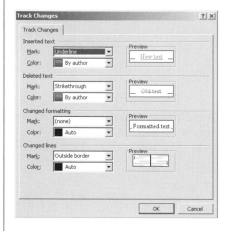

Notice that you can specify the color of the inserted and deleted text as By Author. If different reviewers make changes in your document, a different color is used for each reviewer. If you want all the changes to be displayed in the same color, select the color instead of By Author.

To see which author made the change in your text, hold the cursor over the change for a second to display the ScreenTip showing the author, date and time, and type of change.

in a colored font. Anything someone deletes will be crossed out and be in a colored font.

When you get your peer-reviewed document back, open it, and then select Tools | Track Changes on the menu bar, and then Accept or Reject Changes. The Find arrow will take you to the locations where changes were made, and then you can accept or reject them. Alternatively, go to View | Toolbars, and turn on the Reviewing toolbar by clicking the check box. The commands on the toolbar are nearly the same as those in the Accept or Reject Changes dialog box (Accept all and Reject all are missing).

Make sure to print out the revised version of your paper, and proofread the hard copy. Some mistakes are more easily identified on paper, and it is always better for you, rather than your instructor, to find them.

4. Special Functions in Scientific Papers

Scientific papers possess features that are usually not found in papers written in the humanities. These include:

Greek letters and mathematical symbols
Superscripts and subscripts
Italics to indicate the scientific name of an organism

Tables inserted in the document
Figures imported from another program such as Excel
Equations
Space left to add hand-drawn sketches later

As in MLA style, hanging indents are used to separate references in name-year format. Use the following Quick Reference table (Table A1.3) if you are already familiar with these features.

4.1 Greek Letters and Mathematical Symbols

Some instructors may allow you to write Greek letters and mathematical symbols in by hand. The problem with this approach is that it is easy to

TABLE A1.3 Quick reference for features used in scientific papers

FEATURE	QUICK REFERENCE
Greek letters	Insert	Symbol	Font -» (normal text) or Symbol
Mathematical symbols	Insert	Symbol	Font -» (normal text) or Symbol
Superscripts and subscripts	Format	Font -» Superscript or Subscript
Italics	*I* command from Formatting toolbar
Customize toolbar	Tools	Customize	Commands tab	Categories: Format	Scroll to desired command	Click command and drag it to desired location on Formatting toolbar
Shortcut keys	Insert	Symbol	Font -» (normal text) or Symbol	Select character	Shortcut Key	Define combination of keystrokes	Assign
AutoCorrect	Tools	AutoCorrect	Replace ~ With ~
Tables inserted in the document	Table	Insert Table
Figures imported from another program such as Excel	See Appendix 2
Equations	Insert	Object	Microsoft Equation 3.0
Measuring space	Word processor's ruler
Hanging indent to list references	Format	Paragraph (Indents and Spacing tab)	Indentation: Special	Hanging

miss one of the places where these characters should be inserted in your text, and this oversight may cost you points on your grade. Furthermore, handwritten characters make the paper look unprofessional.

Take some time to locate the symbols you need before you write your paper.

1. Click Insert | Symbol on the menu bar.

2. Select Font: (normal text) and browse the available characters in this character set (Figure A1.22). You will find commonly used scientific symbols such as:

Symbol	Meaning
°	degree(s)
±	plus or minus
μ	micron (10^{-6} m)
‰	parts per thousand
f	function (calculus)

In the normal text character set, you will also find characters with diacritical marks used in European languages other than English. Some common letters with diacritical marks are:

Name	Example
Accent	À, Á, È, É, Ì, Í, Ò, Ó, Ù, Ú, à, á, è, é, ò, ó, ì, í, ù, ú

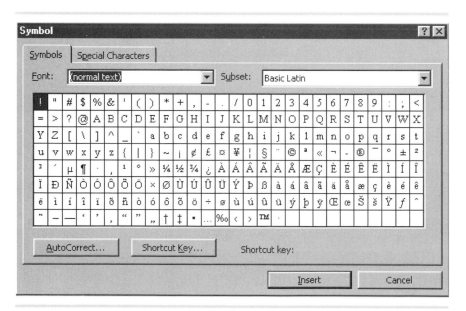

Figure A1.22 Symbol dialog box showing characters under Font: (normal text)

Circumflex	Â, Ê, Î, Ô, Û, â, ê, î, ô, û
Cedilla	Ç, ç
Tilde	Ã, Ñ, Õ, ã, ñ, õ
Umlaut	Ä, Ë, Ï, Ö, Ü, Ÿ, ä, ë, ï, ö, ü, ÿ

When you change the Font to Symbol (Figure A1.23), you will find upper- and lowercase Greek letters and other symbols used in mathematics and chemistry such as:

Symbol	Meaning
$\leq, \geq$	is less than or equal to, is greater than or equal to
∞	infinity
•	raised period (centered dot); multiplication symbol in equations; connection for adducts in a chemical formula
$\leftrightarrow, \leftarrow, \uparrow, \rightarrow, \downarrow$	arrows used to write chemical reactions
∂	partial derivative (calculus)

Once you have found the letter or symbol you need, simply click it, and select the Insert button at the bottom of the dialog box. If you use this letter or symbol frequently, you can make a shortcut key to save time (see Section 4.5).

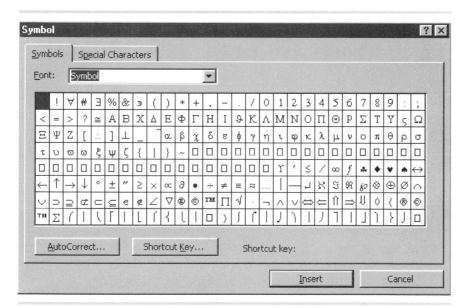

Figure A1.23 Symbol dialog box showing characters under Font: Symbol

4.2 Superscripts and Subscripts

Some expressions used in the natural sciences have superscripted or subscripted characters. It is *not correct* to write these characters on the same line as the rest of the text. Similarly, when using scientific notation, exponents are always superscripted. It is *not acceptable* to write exponents preceded by a caret (^) or by an uppercase E to designate superscript.

> **RIGHT:** 2×10^{-3} Exponent is superscripted
>
> **WRONG:** $2 \times 10{-}3$ or $2 \times 10{\wedge}{-}3$ or $2 \times 10E{-}3$

To superscript or subscript text:

1. Click Format | Font on the menu bar (Figure A1.24).

2. Check the appropriate box under Effects in the Font dialog box.

3. Click OK. The text you type after this formatting change will all be super- or subscripted until you go back to the Font dialog box and deselect that effect.

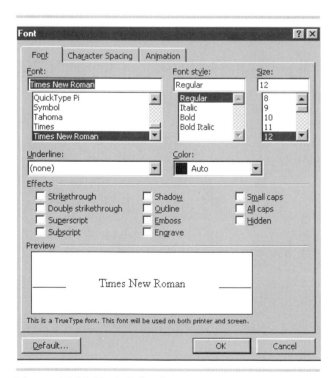

Figure A1.24 Font dialog box

You can also sub- or superscript previously typed text by highlighting the relevant character(s), clicking Format | Font, and selecting the appropriate box under Effects. To highlight text, hold down the left mouse button, and drag it across the text from the first character to the last. If you use super- and subscripts frequently, you can add these commands to the Formatting toolbar (see Section 4.4).

4.3 Italics

By convention, scientific names given to organisms (viruses, bacteria, plants, and animals) are in Latin and are written in italics. Select the *I* command from the Formatting toolbar before you write the scientific name of the organism, and then deselect it when you are finished. You can also highlight the name after you have typed it, and then select the *I* command.

If you have to type certain scientific names frequently, use Auto-Correct (see Section 4.6) either to replace the name with a simple keystroke combination or to have it italicized automatically.

4.4 Customizing Your Toolbars

After you have used Word for a while, you may notice that you never use some of the buttons on the toolbars, but you do frequently use commands that are buried deep in the submenus.

Adding commands to toolbars. You can add frequently used commands, such as super- and subscripts, to the Formatting toolbar. To do this:

1. Click Tools | Customize on the menu bar. The Customize dialog box has three tabs: Toolbars, Commands, and Options.

2. Select the Commands tab (Figure A1.25). The left side shows different Categories, such as File, Edit, View and the other categories on the menu bar. The right side shows the Commands available for the category selected on the left. These commands are generally the options available in the pulldown menus for the respective menu bar categories.

3. To add the superscript or subscript command to the Formatting toolbar, click the Format category (left side of dialog box), and scroll down to the Superscript or Subscript command (right side of dialog box).

4. Click superscript or subscript, and, holding down the left mouse button, drag the command to the desired location on the Formatting toolbar. Instead of having to go through Format

Figure A1.25 Customize dialog box, Commands tab

| Font | Select Superscript or Subscript box, you only have to
click on the appropriate toolbar command.

If you use Word a lot, it may be worth your while to browse through the categories and commands in the Customize dialog box. Make frequently used commands more accessible by adding them to the toolbar or menu bar, and remove commands that you never use.

Removing buttons from toolbars. To remove a button from a toolbar, hold down the Alt key, click the unneeded button, and drag it away from the menus and toolbars. An *x* appears on the icon to show that the command will be removed. If you change your mind later, you can add it back, following the instructions in the previous section.

4.5 Shortcut Keys

It is very time consuming to go through the "Insert | Symbol | Choose Font | Click symbol | Click Insert" routine every time you want to insert a degree sign in your paper. It is worth your while to make shortcut keys for frequently used symbols. To do this, go to Insert | Symbol | Font, and then click the desired symbol. Notice the Shortcut Key box at the bottom of the Symbol dialog box (see Figure A1.22). When you click this box, you have the opportunity to define a combination of keystrokes for that particular symbol (Figure A1.26). The combination should be easy to remember and is usually Ctrl, Alt, Ctrl+Shift, or Ctrl+Alt plus some other character.

Let's say you want to make a shortcut key for the degree sign. The degree sign looks like a lowercase *o*, so you might use Alt + o. You type Alt + o into the Press new shortcut key box, and Word notifies you either that this combination is unassigned (as in Figure A1.26) or that it has already been assigned to another command.

Note: even though you typed Alt and lowercase o, the shortcut appears as Alt and uppercase *O* in the box. If you had typed Alt and uppercase *O*, this would have appeared as Alt + Shift + O.

If the shortcut is already assigned to a different command, determine if you use that command very often. If you don't, you can remove the shortcut from the seldom-used command and assign it to the symbol you want. To do this:

1. Select Assign from the buttons on the right side of the dialog box (see Figure A1.26). Alt + o will then be listed under Current keys. If Word assigned a different shortcut key to the degree sign, this will also be displayed under Current keys (in this example, Ctrl + @, Space).

2. Click the previously assigned shortcut key, and then select Remove from the boxes on the right side of the dialog box.

3. Close the Customize Keyboard dialog box.

4. Close the Symbol dialog box.

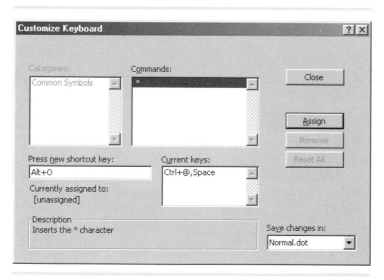

Figure A1.26 Customize Keyboard dialog box

Next time you have to write the symbol for "degrees Celsius," simply type Alt + o followed by uppercase C to get: °C.

4.6 AutoCorrect

AutoCorrect is useful for more than just correcting spelling mistakes. Four possible applications in scientific writing are:

- To replace long chemical names with a simple keystroke combination
- To insert symbols or special characters using just one keyboard key
- To italicize scientific names of organisms automatically
- To insert expressions with sub- or superscripts automatically

Replacing long chemical names with a simple keystroke combination. It is tedious to type units and long chemical names that occur frequently in your text. To save yourself some keystrokes, you can program AutoCorrect to replace an expression that takes a long time to type with a simple keystroke combination. You must choose the simple keystroke combination judiciously, however, because every time you type these keystrokes, Word will replace them with what you programmed in AutoCorrect.

Let's say "beta-galactosidase" is a word you have to type frequently. You choose "bg" to designate "beta-galactosidase." To program Auto-Correct for this entry:

1. Select Tools | AutoCorrect from the menu bar. This opens the AutoCorrect dialog box.
2. Type "bg" in the Replace text box.
3. Type "beta-galactosidase" in the With text box.
4. Click Add button.
5. Click OK.

Inserting symbols or special characters using just one keyboard key. If you want to use AutoCorrect to insert symbols or special characters, the procedure is as follows:

1. Select Insert | Symbol from the menu bar. This opens the Symbol dialog box (see Figure A1.22).
2. Change the font if the symbol you want to use is not in the (normal text) font.
3. Click the symbol.

4. Click the AutoCorrect button at the bottom of the dialog box. The symbol will already be entered into the With text box.

5. Type the characters you want to replace with the symbol in the Replace text box.

6. Click Add button.

7. Click OK.

Italicizing scientific names of organisms automatically. If you are going to be typing the scientific name of an organism repeatedly, you can insure that it always appears in italics without manually imposing italics each time. To do this, follow these steps:

1. First type the scientific name of the organism in your text (e.g., **Aphanizomenon flos-aquae**).

2. Italicize the name by highlighting it (to highlight text, hold down the left mouse button, and drag it across the text from the first character to the last) and then clicking the Italicize icon (*I*) on the Formatting toolbar (the text then becomes: *Aphanizomenon flos-aquae*).

3. With the name still highlighted, click Tools | AutoCorrect on the menu bar.

4. In the AutoCorrect tab, you will notice Aphanizomenon already entered in the Replace text box. Click "Formatted text" next to the With text box. The italicized name is automatically entered.

5. Click the Add button.

6. Click OK to close the window.

Inserting expressions with sub- and superscripts. The principle of using AutoCorrect to insert expressions with sub- and superscripts is the same as that for italicizing scientific names.

1. First type the expression without sub- or superscripts (e.g., Vmax).

2. Highlight the characters to be sub- or superscripted, and then carry out the command, as described in Section 4.2 or 4.4 (the expression then becomes V_{max}).

3. Highlight the entire expression, and click Tools | AutoCorrect.

4. In the AutoCorrect tab, you will notice "Vmax" already entered in the Replace text box. Click "Formatted text" next to the With text box. The correctly subscripted expression is automatically entered.

5. Click the Add button.
6. Click OK to close the window.

4.7 Tables Inserted in the Document

Depending on the guidelines of the scientific journal, tables may either be attached on separate pages at the end of the paper or may be inserted as soon as feasible following the text where the table is first mentioned. Ask your instructor if he or she has a preference. Most people find it easier to refer to a table that is in close proximity to the text.

Word makes it easy to create tables to your specifications and insert them wherever you want in the document (put the cursor at the desired location before you select Table | Insert Table). See Section 2.12 for details.

4.8 Figures Imported from Another Program Such as Excel

As with tables, figures may either be attached on separate pages at the end of the paper or may be inserted as soon as feasible following the text where the figure is first mentioned. Ask your instructor if he or she has a preference. Most people find it easier to refer to a figure that is in close proximity to the text.

The general idea behind importing figures from another program into Word is:

1. Click the desired graphic.
2. Select Copy from the Standard toolbar (or the menu bar) in the graphing program.
3. Position the cursor at the location in the Word document where you would like to import the graphic.
4. Select Paste from the Standard toolbar in Word.

To import a graph (figure) from Excel, follow these steps:

1. First you must make the graph in Excel (see Appendix 2). **Make all formatting changes in Excel,** as it is not possible to make any changes (except to the size) once the graph has been imported into your Word document.
2. When you are satisfied with the graph in Excel, single-click the area outside the axes, but inside the frame (Chart Area, not Plot Area).
3. Still in Excel, click the Copy command in the Standard toolbar.
4. Position the cursor at the desired location in your Word document.

5. Now in Word, click the Paste command in the Standard toolbar.

6. The figure will appear in the Word document.

7. Figures are always numbered and titled **beneath** the figure. The numbers and titles (called collectively the **caption**) may be centered or aligned flush on the left margin. Arabic numbers are used, and the figures are numbered consecutively in the order they are described in the text.

8. Figure titles should be informative and consist of a precise noun phrase (not a complete sentence). Titles that merely state the y-axis legend versus the x-axis legend are *not acceptable*.

4.9 Equations

It is unlikely that you will need to type complex mathematical equations in a scientific paper for an introductory biology course. If you are interested in doing so, however, you can open Microsoft Equation Editor by selecting Insert, and then Object. Select Microsoft Equation 3.0 as the object type. Several references in the Bibliography provide instructions.

4.10 Leaving Space for Sketches

In some laboratory exercises, you may be asked to sketch cells, tissue sections, or specimens that you viewed under the microscope. By convention, these sketches are called "figures" and require a figure caption (number and title) beneath the figure.

It is difficult to judge on the computer screen how much space to leave for your sketches. You don't want to leave too small a space, because your sketches should be large, clear, and with some detail. On the other hand, you don't want to leave too much space, because then it looks like something is missing.

To measure space precisely, invest in a word processor's ruler, which you can buy at any good office supply store. This ruler has a scale that is divided into tenths of an inch, the same units used to tell you the distance of the cursor from the top of the page (see Status bar at the bottom of the screen).

One technique to determine how much space to leave for your sketch is shown in Figure A1.27.

Step-by-step instructions for this method are as follows:

1. First make a rough draft of your sketch so that you have an idea of the space requirement (Figure A1.27A).

2. Type the first draft of your scientific paper (Figure A1.27B).

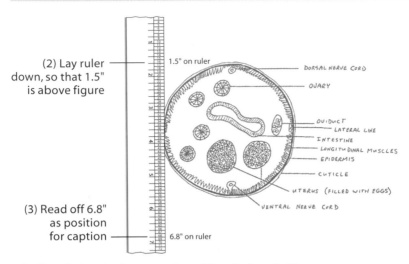

(2) Lay ruler down, so that 1.5" is above figure

1.5" on ruler

DORSAL NERVE CORD

OVARY

OVIDUCT
LATERAL LINE
INTESTINE
LONGITUDINAL MUSCLES
EPIDERMIS

CUTICLE

UTERUS (FILLED WITH EGGS)

VENTRAL NERVE CORD

(3) Read off 6.8" as position for caption

6.8" on ruler

A. Rough sketch of cross section of female *Ascaris*, 40x

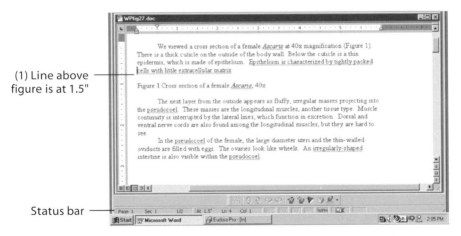

(1) Line above figure is at 1.5"

Status bar

B. Computer screen displaying text into which you plan to insert a sketch

Figure A1.27 A technique for measuring space for hand-drawn sketches. After making a rough sketch on paper (A), and typing your document on the computer (B), (1) position the cursor just above the insertion location in the document. Look at the status bar at the bottom of the screen to determine this distance from the top of the page (in inches). In this example, the distance is 1.5". (2) Lay the word processor's ruler on your sketch so that 1.5" is about 0.5" above the top of your sketch. (3) Read the distance (in inches) at the point about 0.5" below the bottom of your sketch. In this example, that distance is 6.8". (4) Insert Returns in your document until the cursor is at the distance you determined in (3) (see status bar). Finally, type the figure caption on this line.

3. Determine where you want to insert the sketch in your text. It is preferable to insert the figure as soon as possible after the paragraph where it is first described.

4. Position the cursor on the last line of the paragraph in which the figure is described. Note the distance of this line from the top of the page (in inches), as indicated on the status bar at the bottom of the screen. In Figure A1.27B, this is 1.5".

5. Lay the ruler on your sketch so that the mark for the line distance you determined in step 4 is placed just above the top of your sketch.

6. Note the line distance on the ruler just beneath your sketch. In Figure A1.27A, this is 6.8". This number represents the line distance at which you will type the caption for your sketch. If this number is greater than roughly 9.6", you will have to place your sketch on the next page, because there is not enough room for it on the current page.

7. If there is not enough room for the sketch at the bottom of the page, use your judgment to decide whether or not to type more text after the paragraph in which your sketch was first mentioned. If you decide not to type more text, insert a page break, and continue with the text after the caption for your sketch.

4.11 Hanging Indent to List References in the Name-Year System

The Council of Biology Editors manual (6th ed.) recommends the following two formats for listing references:

- Citation-Sequence System
- Name-Year System

If you use the Citation-Sequence system, the references are numbered according to order of citation. See Section 2.13 for instructions on how to make numbered lists.

If you use the Name-Year system, the references are listed in alphabetical order according to the first author's last name. The first line of the reference begins on the left margin, and the subsequent lines are indented. This style of indent is called a **hanging indent**. References typed with hanging indent format look like this:

Murata T, Kadota A, Wada M. 1997. Effects of blue light on cell elongation and microtubule orientation in dark-grown gametophytes of *Ceratopteris richardii*. Plant and Cell Physiology 38(2): 201–209

There are several ways to apply hanging indent format to the list of references. Perhaps the easiest is to type each reference so that it is aligned on the left margin, with an Enter (Return) at the end of each reference. When you are finished, highlight all the references (to highlight text, hold down the left mouse button, and drag it across the text from the first character to the last), and select Format | Paragraph on the menu bar. In the Paragraph dialog box (see Figure A1.10), scroll down to Hanging in the Indentation: Special: box. You can set the indentation to 0.25" or 0.5" in the By: box.

MAKING XY GRAPHS IN MICROSOFT EXCEL

Graphic illustrations are characteristic of scientific and technical writing. They help the reader to visualize the data and to see relationships between variables. They organize information that would otherwise involve lengthy descriptions that would be difficult to interpret.

Graphic illustrations fall into two categories: tables and figures. A table is defined by Webster's dictionary as "a systematic arrangement of data usually in rows and columns for ready reference." **A figure is anything that is not a table.** Thus, line graphs, bar charts, drawings and photographs are always called figures.

Years ago, authors used drawing instruments, specially lined graph paper, and ink to make figures for published papers. No mistakes were permitted. If a different size figure was required, the entire figure had to be redrawn. The advent of computer plotting software in the mid-1980s, however, has made hand-drawn graphs practically obsolete. By entering the data into the graphics program in a prescribed manner and selecting the desired form for the graph, you can have the computer draw the graph for you.

The most common types of graphs are line graphs, bar graphs, and pie charts. The type of graph you choose should be based on which one best depicts and emphasizes the trends shown by your data. Some guidelines for helping you choose an appropriate type of graph are given in the section "Types of Graphs".

In biology, XY graphs are probably the most frequently used, which is why they are the focus of this appendix. With practice, you can make these graphs exactly to your specifications using Microsoft Excel 97 (and higher versions of Excel). Excel is a good plotting program for novices for the following reasons:

- Data input and subsequent plotting of these data is relatively straightforward in Excel.

- Excel is readily available and is included in the Microsoft Office suite of computer software.
- If your school has Excel on its computers, it's likely that you can get assistance from a staff member in your school's computer services department.

The time you invest now in learning to plot data on the computer will be invaluable in your upper-level courses and later in your career. You may eventually switch to a higher-powered plotting program such as SigmaPlot®, but the experience gained by working in Excel should make this transition easier.

The format described in this appendix follows the CBE Manual (1994). Instructions to "single-click" or "double-click" refer to clicking the **left** mouse button, unless otherwise noted.

Excel commands are distinguished in this appendix with a vertical bar separating commands from subcommands. For example, **Tools | Customize | Options** means "Open the Tools menu, then select Customize, and then select the Options command."

Types of Graphs

Frequently encountered graphs are line graphs, bar graphs, and pie charts. But the first question you should ask yourself is, "Is a graph appropriate?"

- If you can state the results in one sentence, **then neither a graph nor a table** is needed.
- If the numbers themselves are more important than the trend shown by the numbers, then a **table** should be used.
- If you have ruled out these two cases, then nothing beats a graph to make a visual impression.

Pie Charts

A pie chart is used to show data as a percentage of the total data. For example, if you were doing a survey of insects found in your backyard, a pie chart would be effective in showing the percentage of each kind of insect out of all the insects sampled (Figure A2.1).

Bar Graphs

A bar graph allows you to compare individual sets of data when these data are **non-numerical** or **discontinuous**. For example, if you wanted

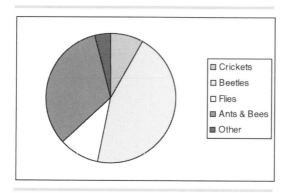

Figure A2.1 Composition of insects in backyard survey

to compare the final height of the same species of plant treated with four different nutrient solutions (Figure A2.2), each bar could be used to represent the plants treated with one nutrient solution. The data plotted on the *x*-axis are non-numerical, because different nutrient solutions (not numbers) are being compared. The data are discontinuous, because the individual solutions have no effect on the growth of the plants treated with the other nutrient solutions.

Excel distinguishes between bar charts and column charts. Figure A2.2 is an example of a column chart, where the categories are organized horizontally and the numerical values vertically. Figure A2.3 is an example of a bar chart, with the categories organized vertically and the numerical values horizontally. According to Excel, bar charts are preferred when the emphasis is on comparing values rather than their variation over time.

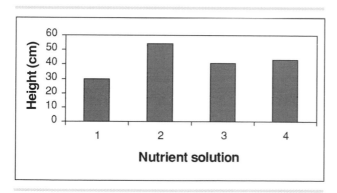

Figure A2.2 Final height of corn plants after 4-week treatment with different nutrient solutions. This figure is an example of a column chart made by Excel.

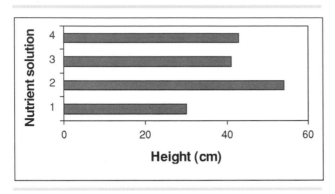

Figure A2.3 Final height of corn plants after 4-week treatment with different nutrient solutions. This figure is an example of a bar chart made by Excel.

Line Graphs

Line graphs (or XY graphs) are perhaps the most frequently used type of graph in biology. Line graphs are used to display a trend or an important relationship between one or more variables. The data displayed in line graphs are **continuous** and **numerical**; this is the main difference between line graphs and bar graphs.

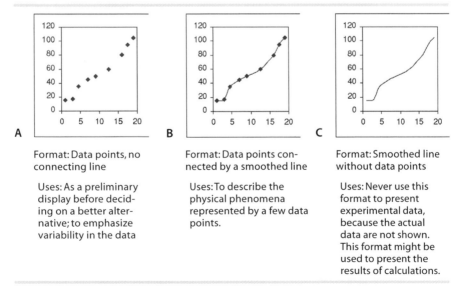

A — Format: Data points, no connecting line

Uses: As a preliminary display before deciding on a better alternative; to emphasize variability in the data

B — Format: Data points connected by a smoothed line

Uses: To describe the physical phenomena represented by a few data points.

C — Format: Smoothed line without data points

Uses: Never use this format to present experimental data, because the actual data are not shown. This format might be used to present the results of calculations.

Figure A2.4. Different formats for XY graphs

By convention, the independent variable (the one the scientist manipulates) is plotted on the *x*-axis, and the dependent variable (the one that changes in response to the independent variable) is plotted on the *y*-axis.

Excel offers a number of alternatives when it comes to plotting line graphs; however, the most obvious one, Line Graph, should NOT be used to plot scientific data. The Line Graph alternative makes all the intervals on the *x*-axis equal, even when the data are not in equal intervals. Choose XY Scatter instead.

There are a number of formats for plotting data on an XY graph (Figure A2.4). Situations in which each format might be used are given.

The Microsoft Excel Screen

Double-click the Microsoft Excel icon on your desktop screen to open an Excel worksheet (spreadsheet). A worksheet consists of thousands of cells arranged in rows and columns, with menus and toolbars located above the cells at the top of the screen display (Figure A2.5). The columns have letter headings (A, B, C, etc.) and the rows have numerical headings (1, 2, 3, etc.).

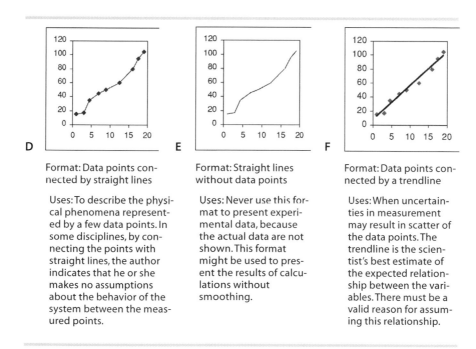

D	**E**	**F**
Format: Data points connected by straight lines	Format: Straight lines without data points	Format: Data points connected by a trendline
Uses: To describe the physical phenomena represented by a few data points. In some disciplines, by connecting the points with straight lines, the author indicates that he or she makes no assumptions about the behavior of the system between the measured points.	Uses: Never use this format to present experimental data, because the actual data are not shown. This format might be used to present the results of calculations without smoothing.	Uses: When uncertainties in measurement may result in scatter of the data points. The trendline is the scientist's best estimate of the expected relationship between the variables. There must be a valid reason for assuming this relationship.

Figure A2.4. *continued*

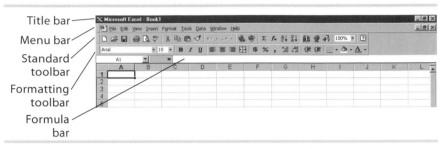

Title bar
Menu bar
Standard toolbar
Formatting toolbar
Formula bar

Figure A2.5 Screen display of Microsoft Excel 97 worksheet

Title bar

The title of the worksheet is given at the very top of the page on a blue background. If you have not yet named the worksheet, the title bar will read "Microsoft Excel—Book 1."

Menu bar

Just below the title bar is the menu bar. The menu bar consists of nine commands that, when selected, are subdivided into further commands listed in drop-down menus. For example, when you click File, the drop-down menu shown in Figure A2.6 is displayed. Single-click the left mouse button on the desired command to select that command.

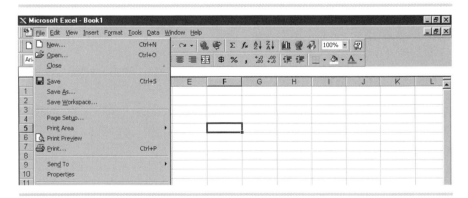

Figure A2.6 Drop-down menu for File command

Excel 2000 Update

Standard and Formatting toolbars share one line, instead of being on separate lines. You can change this to two lines (if desired) in Tools | Customize | Options.

Toolbars

If your Excel program has been set up to display them, the toolbars (Standard and Formatting) are displayed just below the menu bar. Toolbars contain icons (pictures) that are shortcuts to menu commands.

If the Standard and Formatting toolbars are not displayed, go to the menu bar and select View | Toolbars to see if there is a checkmark next to Standard and Formatting. If there is no checkmark, go to the last item on the Toolbars drop-down menu and click Customize. Click the boxes next to Standard and Formatting, then click Close.

Toolbars contain icons (pictures) that are shortcuts to menu commands. If you place the cursor (arrow) over a particular icon (do not click any mouse buttons) on either toolbar, a description of that icon will appear. For example, if you place the cursor on the fourth icon from the left on the Standard toolbar (Figure A2.7), the Print command, with the default printer in parentheses, will appear in a pop-up label. The Print command is also found on the menu bar under File | Print.

Note: Shortcuts offer only one option from the corresponding menu command. For example, if you click the printer icon on the standard toolbar, the entire document will be printed on the default printer. If you want to print single pages or use a different printer, you must go to the menu bar and select the desired options from the File | Print drop-down menu.

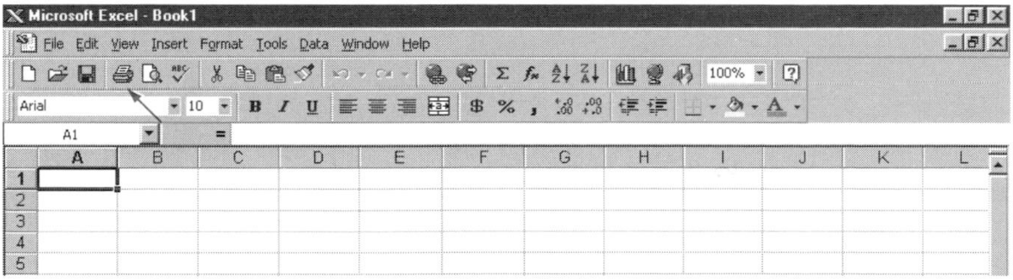

Figure A2.7 Selecting the printer icon from the Standard toolbar

Formula bar

The formula bar is located just below the toolbars. It is used to enter formulas for performing calculations on the data entered in the spreadsheet. It is also used to edit the contents of a cell.

Entering Data in Spreadsheet

Before you enter data in an Excel worksheet, you must first have a clear idea of what your XY graph should look like. Which parameter should be plotted on the x-axis and which one on the y-axis? By convention, the x-axis of the graph shows the independent variable. The independent variable is the one that the scientist manipulated during the experiment. The y-axis of the graph shows the dependent variable, the one that changes in response to changes in the independent variable.

Let's say you did an experiment in which you determined how the rate of activity of an enzyme changed when you varied the temperature. Because you manipulated the temperature in the experiment, temperature is the independent variable that should be plotted on the x-axis. Enzyme activity is plotted on the y-axis, because it changes in response to changes in the temperature.

On the Excel spreadsheet, **Column A** is used to enter the data for the **x-axis,** whereas subsequent columns are used for data for the y-axis (Table A2.1). In the previous example, temperature data would be plotted in Column A and the corresponding enzyme rate data would be plotted in Column B.

Enter the data as in Table A2.2. For a simple graph with only one data set, it is not necessary to provide column headings.

Note: Make sure you enter the data for the x- and y-axes in *neighboring columns.* If the data for the two axes are separated by one or more columns, Chart Wizard does not make the graph properly.

TABLE A2.1 Portion of Excel 97 spreadsheet

	A	B	C	D	E
1					
2					
3					
4					
5					

TABLE A2.2 Sample data for effect of temperature on rate of enzyme activity. Temperatures (°C) are entered in Column A, rates of activity (sec^{-1}) in Column B.

	A	B
1	4	0.039
2	5	0.073
3	3	0.077
4	0	0.096
5	7	0.082
6	0	0.04
7	0	0.007
8	0	0
9	0	0

Terminology

The terminology that Excel uses for graphs is different from what biologists use. You have already learned that Excel calls a spreadsheet a worksheet. Some other terms are translated in Table A2.3.

These Excel instructions use the terms biologists use, rather than those chosen by Excel. It is helpful to know the Excel terminology when you use the Help menus, however.

TABLE A2.3 Excel-specific terminology

EXCEL TERM	DESCRIPTION
Chart	Graph
Category axis	x-axis
Value axis	y-axis
Data series	Set of related data points
Plot area	Area of the graph inside the axes
Chart area	Area outside the axes but inside the frame
Legend	Legend or key

Using Chart Wizard to Plot the Data

1. In the spreadsheet, highlight the data to be plotted using the mouse. To do this, single-click the left mouse button on the first cell, and keep holding the button down as you drag the cursor over the rest of the cells that contain the data to be plotted. When you release the mouse button, the cells you selected are highlighted in blue.

2. Select Chart Wizard from the Standard toolbar. The icon is a colored bar graph (Figure A2.8).

Using four sequential dialog windows, Chart Wizard walks you through most of the steps needed to make a graph.

3. Chart Wizard Step 1 of 4—Chart Type allows you to select the type of plot (Figure A2.9).

   For an explanation of each chart type, click the ? at the bottom left of the dialog window.
   - To make a line graph, select XY (Scatter). Do *not* select Line, because this option spaces the *x*-axis values at equal intervals, instead of according to the intervals of the data.
   - For Chart sub-type, you must decide whether the points should be connected or not. If the data show the relationship between an independent variable and a dependent variable (e.g., effect of temperature on rate of enzyme activity), then select Data points connected by smoothed or straight lines. Usually your instructor wants to see the data points, so do *not* select the options in which Excel plots the curve without displaying the data points. If you know that the relationship between the independent and dependent variables is supposed to be linear (i.e., you are plotting a standard curve–also called a regression line or trendline), then select Scatter with no points connected.
   - Click Next.

4. Chart Wizard Step 2 of 4—Chart Source Data tells you what range you highlighted on the spreadsheet. The data series should be in Columns. Click Next.

Figure A2.8 Selecting Chart Wizard from the Standard toolbar

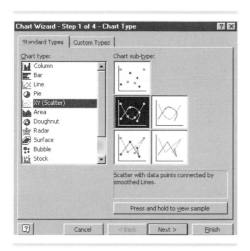

Figure A2.9 Selecting a type of graph in Chart Wizard, Step 1 of 4. Current selection is XY Scatter with data points connected by smoothed lines.

5. Chart Wizard Step 3 of 4—Chart Options shows you five tabs: Titles, Axes, Gridlines, Legend, and Data Labels (Figure A2.10).

📁 **Titles**

Chart Title: Leave blank. By doing so, Excel will make a larger graph for you to import into your text document later. In the text document, type the figure caption *under* the figure.

Value (X) axis: Enter the *x*-axis label with the units in parentheses. For the data in Table A2.2, the *x*-axis label

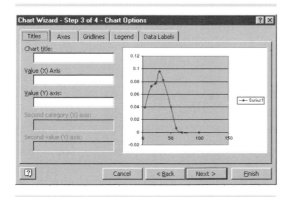

Figure A2.10 Dialog window for entering graph title (Chart title), *x*-axis label, and *y*-axis label. Leave Chart title blank.

would be: Temperature (°C). For the time being, don't worry about the superscripted degree sign. Simply type lowercase o.

Value (Y) axis: Enter the y-axis label with the units in parentheses. For the data in Table A2.2, the y-axis label would be: Rate (sec^{-1}). For the time being, don't worry about the superscripted exponent. Simply type "-1".

📁 **Axes**

No changes.

📁 **Gridlines**

There should be *no* gridlines on the figure. If the default is Value (Y) axis: Major gridlines, click the check mark to remove this selection.

📁 **Legend**

No legend (key) is required when there is only one line on the graph. If the default is Show legend, click the check mark to remove the legend.

📁 **Data labels**

None.

When you have finished with these five tabs, click Next.

6. Chart Wizard Step 4 of 4—Chart Location asks you if you want the figure to be printed as a new sheet or as an object in the current sheet.
   - By convention, figures are placed as close as possible to the location in the text document where they are first described. Because you have to import the graph into a text document anyway, it does not really matter whether you choose As object in Sheet 1 or As new sheet. Selecting As object in Sheet 1 has the advantage that graph size is uniform when the object is imported into the text document. You can, however, adjust the size of the graph in the text document with either option.
   - If you plan to attach the figure on a separate sheet at the end of the document (may be acceptable, but is not preferred), then select As new sheet.
   - Then click Finish. Your graph should look like Figure A2.11.
7. Inspect the graph carefully. Look for proper units and spacing on the x- and y-axes, for appropriate labeling of the axes, and

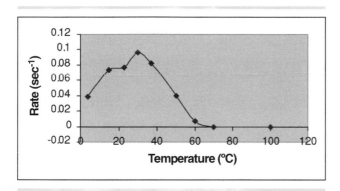

Figure A2.11 First draft of Table A2.2 data plotted by Chart Wizard

for expected trends of the line itself. If you notice that you entered a datum incorrectly in the spreadsheet, simply change it, and the correction will also be made in the graph.

Modifying Graphs

Graphs produced by Chart Wizard can be modified to meet an instructor's or a journal's specifications. To modify a graph, first activate it by single-clicking on the plot area or chart area (Figure A2.12). Then select the Chart command from the menu bar. You can return to any of the four steps of Chart Wizard (Chart Type, Source Data, Chart Options, and

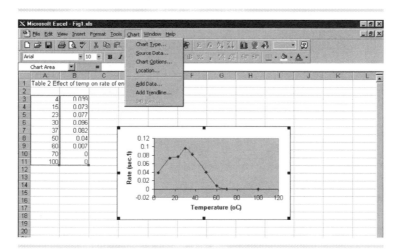

Figure A2.12 Activating a graph and selecting the Chart command from the menu bar

Location), or you can select Add Data (type in range of new data to be plotted from spreadsheet) or Add Trendline.

Note: The Chart command will not be displayed on the menu bar unless you first activate the graph (click inside the chart or plot area).

Chart Type

If you want to see what the data look like plotted in another form, select Chart | Chart Type. Select a different type of graph, and select Press and hold to view sample. Check with your instructor regarding what chart type is appropriate.

Source Data

If you want to change the range of the cells used to plot the graph, you can do so in this dialog window. This might be appropriate if you added or deleted data values in the spreadsheet.

Chart Options

This allows you to change the x- and y-axis labels and delete or display gridlines and legends. The Axes tab seems to suggest that the format of the axes can be changed here, but this is misleading. See "Format Axes."

Location

If you expect to insert the figure in a Word document (preferred), place the figure as an object in the worksheet. If you plan to attach the figure on a separate sheet at the end of the document, select As new sheet.

Excel 2000 Update

When you single-click the plot area or chart area, a floating toolbar with Chart commands appears in the center of the screen.

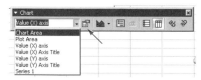

Select any of the Chart Options in the pull-down menu, and then click the Format button to change the format of that particular option. The Chart command also appears in the menu bar (as in Excel 97), where there are more options than provided on the buttons in the floating toolbar. The dialog windows are the same as those in Excel 97.

Add Data

To add data to an existing figure, first enter the new data in cells on the spreadsheet. Then click the chart area of the existing figure, select Chart I Add Data, and enter the range of cells that contain the data to be added to the existing figure.

Add Trendline

A trendline is a line that describes the general tendency of the data points. When Excel plots a linear trendline, it is really plotting a regression line, or best-fit line, with the equation $y = mx + b$.

Trendlines can be used to make standard curves, where the expected relationship between the independent and dependent variables is known. Some examples in biology where standard curves are used include:

- Determining the protein concentration of an unknown sample by measuring its absorbance and comparing the absorbance to that of a series of known protein concentrations (e.g., Biuret method, Bradford method)
- Determining the size of a DNA fragment by comparing the distance it migrated on a gel to the distance migrated by DNA fragments whose sizes are known

Don't confuse a trendline with a smoothed line that connects the data points. If you simply want to show the trend of a set of related data points, select Chart I Chart Type I XY Scatter with data points connected by smoothed lines.

Format Plot Area

Scientific journals are traditionally published in black and white. With this in mind, make the background color (plot area) WHITE for best contrast. To select the background color, double-click the plot area (area inside the axes). In the Format Plot Area dialog box, Area, select None. Set the Border to Automatic.

The background color of the key (legend) should also be white in most cases. Double-click the border for the key to open the Format Legend dialog box. In the Patterns tab, Area, select None (if the default is Automatic: White, the result is the same). Click OK.

When possible, place the key within the plot area. This is done by single-clicking the key (the selection handles around the frame will appear) and dragging it to the desired location.

Format Chart Area

Graphs in scientific journals do not have a frame around them. To remove the frame that Excel automatically creates around the figure, double-click the chart area (area outside the axes, but inside the frame). In the Format Chart Area dialog box, Border, select None. Set the Area to None (if the default is Automatic: White, the result is the same).

Format Axes

Use the criteria in Table A2.4 to check that the axes have been formatted correctly. If any changes need to be made, double-click any number on either the x-axis or the y-axis (depending on which axis needs to be changed). This action opens the Format Axis dialog box (Figure A2.13). Five tabs are displayed: Patterns, Scale, Font, Number, and Alignment. Click the appropriate tab (see Table A2.4) and make the necessary changes.

TABLE A2.4 Checklist for axis format

CORRECT FORMAT	HOW TO ADJUST
Dependent variable on y-axis	See "Entering Data in Spreadsheet"
Independent variable on x-axis	See "Entering Data in Spreadsheet"
Range of values on axes is slightly larger than range of data values being plotted	Scale tab | Min, Max If negative values are not appropriate for the quantity plotted, Min = 0.
Axis numbers are multiples of 2, 5, or 10 whenever possible	Scale tab | Major unit Minor units are not displayed, so enter any value for minor unit.
If scale includes axis numbers less than 1, a zero is required before each decimal point	See "Entering Data in Spreadsheet"
Tick marks are left of the y- axis and below the x-axis	Patterns tab | Major tick mark type: Outside
Tick marks should always be accompanied by a number (no subdivisions between numbered marks)	Patterns tab | Minor tick mark type: None
Numbers should be centered on their respective tick marks, outside the field of the graph	Patterns tab | Tick mark labels: Next to axis
Numbers should be the same size and should read horizontally	Alignment tab | Orientation | Automatic

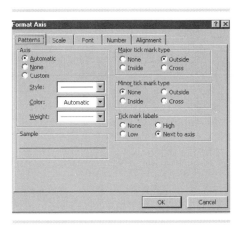

Figure A2.13 Patterns tab inside Format Axis dialog box, opened by double-clicking any number on either the y-axis or the x-axis label

Format Axis Label

Use the criteria in Table A2.5 to check that the x-axis and y-axis labels have been formatted correctly. If any changes need to be made, double-click the appropriate label. This action opens the Format Axis Title dialog box (Figure A2.14). Three tabs are displayed: Patterns, Font, and Alignment. Click the appropriate tab (see Table A2.5) and make the necessary changes.

To **superscript** or **subscript** certain characters in the x- and y-axis labels, or in the key, single-click the text box surrounding the label you wish to modify (the border will then be highlighted). Use the mouse to highlight the character(s) to be super- or subscripted, and then select Format | Select Axis Title on the menu bar. Select either super- or subscript in the Effects category. Click O.K.

Greek characters and other symbols not in the "Normal" character set pose a special challenge. If you know that a special character will be part of the axis label, follow these instructions:

1. For PCs (IBM), click the Start button located in the lower left corner of your screen. This can be done either before or after you use Chart Wizard to plot the data.
2. Select Programs | Accessories | Character Map.
3. Scroll up or down to Font: Symbol.
4. Locate the symbol you want to include on the axis label, and single-click it. Let's use μ (Greek mu) as an example.
5. Now scroll up to Font: Arial. This is the default font for Excel. The character that corresponds to m will be highlighted (m).

TABLE A2.5 Checklist for axis label format

CORRECT FORMAT	HOW TO ADJUST
Capitalize only the first word of the label and any proper nouns	Retype
A word or a phrase that accurately describes the variable	Retype
Centered on the length of the axis	Alignment tab | Text alignment Horizontal: Center Vertical: Center Orientation: 0 degrees
Vertical axis labels read parallel to the y-axis (they should never read vertically downward)	Alignment tab | Text alignment Vertical: Center Orientation: 0 degrees
Units of measurement for the variable are placed in parentheses after the variable	Retype
Font typeface and size	Font tab Font: Arial or another *sans serif* font Font style: Regular or bold Size: 12
Superscripted or subscripted characters	See below
Special characters (e.g., Greek letters)	See below
No borders or background	Patterns tab Border: None Area: None

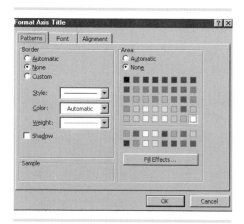

Figure A2.14 Patterns tab inside Format Axis Title dialog box, opened by double-clicking either the *y*-axis or the *x*-axis label

That means that you should type *m* in the axis label wherever you want μ.

6. When you go back to modify the graph, single-click the text box surrounding the label that contains the special symbol. Use the mouse to highlight the *m* that you want to change to μ.

7. On the menu bar, click Format | Selected Axis Title. Under Font, scroll down to highlight Symbol. Then click OK.

Example: Modifying Figure A2.11

The criteria listed in the previous section can now be used to evaluate the graph produced by Chart Wizard (see Figure A2.11). Problems with the graph and how to correct them are summarized in Table A2.6.

TABLE A2.6 Problems with Figure A2.11 and how to remedy them

PROBLEM	REMEDY
Plot area (background) should be white, not gray	Double-click plot area to open Format Plot Area dialog box. Select: Border: Automatic Area: None (or white)
Chart area should not have a frame	Double-click chart area to open Format Chart Area dialog box. Select: Border: None Area: None (or white)
Range of values on axes 1. x-axis: 0 to 120 is 20% larger than needed 2. y-axis: Negative rates don't make sense. Minimum value should be zero.	Double-click any number on the respective axis. 1. In the Scale tab for the x-axis, change Max to 100. 2. In the Scale tab for the y-axis, change Min to 0.
1. Font size: enlarge to 12 pt 2. Axis labels: x axis: degree sign should be superscripted y-axis: −1 should be superscripted	Single-click the respective axis label. 1. Format	Selected Axis Title on the menu bar. In the Font folder, Size: 12. 2. Highlight the character(s) to be superscripted. Then click Format	Selected Axis Title. In the Font folder, Effects: Superscript. Note: The y-axis label will temporarily be turned 90° so that it reads horizontally. When you finish with the formatting, click outside the figure and the label will return to its original orientation.

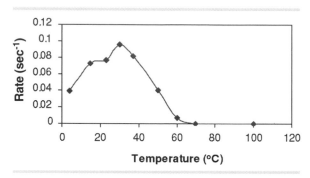

Figure A2.15 Effect of temperature on the rate of enzyme activity

The modified graph (Figure A2.15) is now ready to be copied into your text document (see "Importing Graphs into Microsoft Word").

Importing Graphs into Microsoft Word

In order to give your lab report a professional appearance, insert figures as close as possible to the location in the document where they are first mentioned. This can be done easily by copying the graph you made in Excel and pasting it into a Microsoft Word document (this method also works for WordPerfect documents).

Here are some tips that will help you carry out this task:

1. **Make all modifications to the graph in Excel**. Once it is imported into the Word document, it is no longer possible to make any changes (except to the size of the graph).

2. To copy the graph, single-click the area outside of the axes, but inside the frame (Chart Area, not Plot Area). Selection handles will be displayed around the frame, indicating that the graph is activated.

   Note: If you mistakenly *double-click* the chart area, the Format Chart Area window comes up. Click Cancel to close the window.

   When the graph is activated, select the Copy icon from the Standard toolbar (or Edit | Copy from the menu bar). The frame will then have moving dashes.

3. Move the cursor to the location in the Word document where you want to insert the graph. Select the Paste icon from the Standard toolbar (or Edit | Paste from the menu bar). The graph will be pasted into the document at the position of the

cursor. If you pasted the graph in the wrong place, select Undo Typing from either the Standard toolbar or the menu bar (Edit | Undo Typing). If you notice an error in the graph, make the correction in Excel, and then copy and paste again.

4. Figures are always numbered and titled **beneath** the figure and are always single spaced. Arabic numbers are used, and the figures are numbered consecutively in the order they are discussed in the text.

5. Number and title your figure in the Word document, not in the Chart Title provided in Excel. As explained previously, the omission of the title in Excel results in a larger graph to import into Word. It also allows you to position the title where it belongs: beneath the figure, not above it (which is where Excel places it).

6. Figure titles should be informative and consist of a precise noun phrase (not a complete sentence). Titles that merely state the y-axis label versus the x-axis label are **not** acceptable. Figure titles should **not** begin with "Graph showing…" or a similar description of the figure.

7. There are a number of different acceptable figure caption formats. The most important thing is to be consistent.

   Aligned on left margin:

   > Figure 1 Effect of end-of-day red and far red irradiation on length and width of the third leaf in *Helianthus annuus* "P75"

   First line aligned on left margin, second line indented:

   > Figure 1 Effect of end-of-day red and far red irradiation on length and width of the third leaf in *Helianthus annuus* "P75"

   Centered:

   > Figure 1 Effect of end-of-day red and far red irradiation on length and width of the third leaf in *Helianthus annuus* "P75"

8. In the text of the paper, describe the important results in each figure, and reference the figure parenthetically. For example:

   > Catalase activity reached a peak at 30°C, and no activity was found beyond 70°C (Figure 1).

   Do not write:

   > Figure 1 shows that catalase activity reached a peak at 30°C, and no activity was found beyond 70°C.

The phrasing "Figure 1 shows" should only be used when the figure literally shows a picture of something physical.

The final form of the document with the graph inserted should be something like this:

> The rate of enzyme activity changed with temperature (Figure A2.15). From 4 to 30°C, activity steadily increased to a peak of 0.096 sec^{-1}, and then activity fell with increasing temperature until there was no activity beyond 70°C.

Leave about 1 cm of space below the figure caption to separate it from the subsequent text.

Trendlines

Trendlines are used when uncertainties in measurement may result in scatter of the data points. Trendlines are also called "best-fit" lines, because they represent the scientist's best estimate for the relationship between the dependent and independent variables being studied. Do not use a trendline unless there is a valid reason for assuming a particular mathematical relationship between the variables.

Trendlines are often used to make standard curves in which unknown quantities are determined by interpolation or extrapolation of known quantities. The equation of the trendline allows you to determine the unknown quantities precisely. This eliminates the need for "eyeballing" points off the curve, which is extremely inaccurate by comparison.

Let's say you did an experiment in which you used the Biuret method to determine protein concentration of an unknown sample. To make the standard curve, you measured the absorbance of four known concentrations of bovine serum albumin (BSA). Because of Beer's Law, absorbance is proportional to concentration, so you expect a plot of absorbance vs. protein concentration to be a straight line.

1. Follow the instructions in the section "Entering Data in Spreadsheet" to enter the data in Table A2.7.

2. Follow the instructions in the section "Using Chart Wizard to Plot the Data." This time, in Chart Wizard Step 1 of 4, Chart Sub-type, select the first option, Scatter, **without** connecting the points.

3. The resulting graph should look like Figure A2.16.

TABLE A2.7 Sample data for Biuret standard curve. Protein concentrations (mg/mL) are entered in Column A, absorbance values (at 550 nm) in Column B.

	A	B
1	2	0.10
2	3	0.12
3	5	0.24
4	10	0.40

4. Single-click any point to select the unconnected data set. All the data points should now be highlighted (yellow).

5. Select Chart from the menu bar. Select Add Trendline.

6. There are two tabs under the Add Trendline option: Type and Options. In the Type tab, select Linear (probably the default).

7. Now click the Options tab. Select Display equation on chart by clicking the empty box. This will allow you to plug the absorbance value you measured for the unknown into the equation for y to determine the protein concentration of the unknown (x).

8. Because it may be necessary to determine an unknown concentration below 2 mg/mL, it would be practical if the trendline extended to the origin. Still in the Options tab, set Forecast Backward: to 2 units (or whatever your closest data point to the origin).

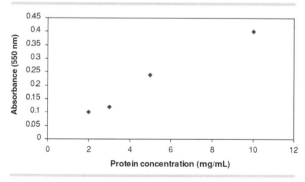

Figure A2.16 Biuret standard curve after using Chart Wizard, changing plot area to white, and changing chart area border to none

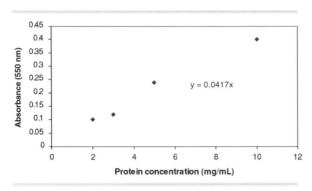

Figure A2.17 Biuret standard curve with trendline extending to the origin

9. Finally, because theoretically the absorbance of a sample that contains 0 mg/mL protein should be 0, select Set the intercept to 0 by clicking on the empty box. Ask your instructor in which cases it is appropriate to do this.

10. Select O.K. The resulting figure should look like Figure A2.17.

   Note: If necessary, the symbols y and x in the equation can be changed to the actual symbols for the variables. Single-click the equation frame and make the necessary changes.

11. Follow the instructions in the section "Importing Graphs into Microsoft Word."

Multiple Lines on One Set of Axes

In some experiments it is desirable to compare the results from a number of different treatments. Plotting the data on one set of axes is often the most efficient way to convey this information. How many lines should you put on one set of axes? The CBE Manual recommends no more than 8, but use common sense. You should be able to follow each line individually, and the graph should not look cluttered.

Because there will be a multitude of lines on the figure, it is critical to identify the individual sets of data by means of a key. Excel will generate a key (legend) for each graph, provided you enter the appropriate titles in the spreadsheet. This can be accomplished by entering a short title in the first row of each of the data columns. These titles will be used by Excel to generate the key.

Let's say you monitored the production of an enzyme called beta-galactosidase in five flasks of *E. coli*, each flask representing a different growth condition. Every 10 min for 1 hr, you measured the absorbance of beta-galactosidase in each culture flask. Time is the independent variable that will be plotted on the x-axis. According to Excel convention, therefore, time should be entered in Column A. The next five columns will contain the absorbance data at each time for the five different flasks. The entries in the data sheet should thus look like Table A2.8.

1. Enter the data into the spreadsheet. Highlight all of the entries (column headings as well as numbers) and select Chart Wizard.

2. Follow the instructions in the section "Using Chart Wizard to Plot the Data."
   - In Chart Wizard Step 1 of 4, Chart Sub-type, select Scatter with data points connected by smoothed lines.
   - In Chart Wizard Step 2 of 4, make sure the data series is in Columns.
   - When you get to Chart Wizard Step 3 of 4, in the Legend tab, this time select Show Legend.
   - In the Titles tab, leave the chart title blank, as before. Enter the axis titles, giving the variable first and the units in parentheses. In the current example,
     x-axis: Time (min)
     y-axis: Absorbance (@420 nm)

3. With so many lines on one set of axes, it is a good idea to enlarge the graph while still in Excel. Pull on a corner of the frame to make this adjustment.

TABLE A2.8 Sample data for β-galactosidase production in *E. coli* grown under five different conditions

	A	B	C	D	E	F
1	Time (min)	Flask 1	Flask 2	Flask 3	Flask 4	Flask 5
2	0	2	0.205	0.1	0.04	0.044
3	10	2	0.265	0.88	0.095	0.174
4	20	2	0.406	1.4	0.2	0.214
5	30	2	0.351	2	0.23	0.3
6	40	2	0.386	2	0.33	0.442
7	50	2	0.56	2	0.35	0.468
8	60	2	0.67	2	0.42	0.91

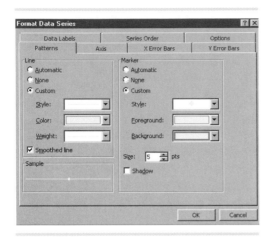

Figure A2.18 Patterns tab inside Format Data Series dialog box, opened by double-clicking either a point or the line connecting the points in a data set

4. If the symbols and the lines are light colored (yellow, light blue, etc.), they will not stand out on a black-and-white hard copy. Improve contrast as follows:
   a. Double-click one of the light-colored lines to display the Format Data Series dialog box (Figure A2.18).
   b. Select the Patterns tab and make the following changes.
      - Line: ☑ Custom: Style (continuous, dashed, etc.), Color (select black), or Weight (default is OK.).
      - Marker: ☑ Custom: Style (circle, square, etc.), Foreground color, or Background color (select black). Remember that most scientific and technical journals are printed in black and white, so the lines should be readily distinguishable by something other than color.
5. The final graph should look something like Figure A2.19.

Multiple Trendlines on One Set of Axes

This is similar to the previous example, except that this time, the lines are trendlines, instead of lines that connect the data points.

Table A2.9 shows sample data for an experiment in which the rate of enzyme activity was measured for different combinations of inhibitor and substrate concentrations. Column A gives the values for the x-axis (1/substrate concentration), while Columns B-E give the points for the

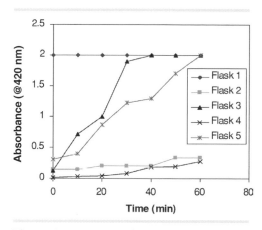

Figure A2.19 Beta-galactosidase production in *E. coli* grown under five different conditions. Flask 1 contains the lacI⁻ strain in glycerol. Flasks 2-5 contain
the lacI⁺ strain. Flasks 2 and 3 contain glycerol medium; flasks 4 and 5 contain
glucose. IPTG was added to flasks 3 and 5 at *t*=0 min.

y-axis (1/rate in seconds). Each column (line on the graph) represents
one inhibitor concentration, which you enter in Row 1 for Columns B–E.
These short titles will be used by Excel to generate the key.

1. Enter the data into the spreadsheet. Highlight all of the entries
   (column headings as well as numbers) and select Chart
   Wizard.

2. Follow the instructions in the section "Using Chart Wizard to
   Plot the Data."
   - In Chart Wizard Step 1 of 4, Chart Sub-type, select Scatter
     without connecting the points.
   - In Chart Wizard Step 2 of 4, make sure the data series is in
     Columns.

TABLE A2.9 Effect of inhibitor concentration on rate of enzyme activity

	A	B	C	D	E
1	1/[H2O2], M	I = 0%	I = 0.0025%	I = 0.025%	I = 0.25%
2	7.14	66.3	60.74	93.34	130.17
3	3.45	39.66	31.3	72.45	114.77
4	0.74	33.77	32.97	48.36	98.76
5	0.34	13.63	14.99	16.81	31.02

- When you get to Chart Wizard Step 3 of 4, in the Legend tab, select Show Legend. You can also change the position of the key as desired (in Figure A2.20, it was placed below the figure).
- In the Titles tab, leave the chart title blank, as before. Enter the axis titles, giving the parameter first and the units in parentheses. In the current example,

  x-axis label: 1/[H2O2], M
  y-axis label: 1/Rate (sec)

3. With so many lines on one set of axes, it is a good idea to enlarge the figure while still in Excel. Pull on a corner of the frame to make this adjustment.

4. If the symbols are light colored (yellow, light blue, etc.), they will not stand out on a black-and-white hard copy. Improve contrast as follows:

   a. Double-click any of the light-colored symbols to display the Format Data Series dialog box (see Figure A2.18).

   b. Select the Patterns tab and make the following changes.

      - Marker: ☑ Custom: Style (circle, square, etc.), Foreground color, or Background color (select black).

      - Line: ☑ NONE. Then close the window by clicking OK.

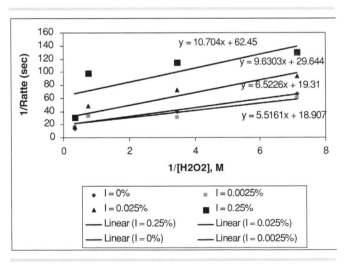

Figure A2.20 Preliminary Lineweaver-Burk plot showing the effect of inhibitor concentration on rate of enzyme activity with different substrate concentrations

5. Now we want to have Excel draw a trendline for each of the data sets. Click one of the symbols to select the data set. All the data points in that set should now be highlighted (yellow).

6. Select Chart | Add Trendline from the menu bar.

7. There are two tabs under the Add Trendline option: Type and Options. In the Type tab, select Linear (probably the default).

8. Now click the Options tab. Select Display equation on chart by clicking the empty box. The equations allow you to determine whether all the lines have a common y-intersept, a common x-intercept or neither. This information is used to determine whether a competitive or noncompetitive inhibitor is involved in the reaction.

9. Select OK. The resulting figure will look something like Figure A2.20.

10. Because the "Linear" entries in the key are meaningless, single-click each of these text boxes (so that the border is highlighted), and then press the Delete key. Superscript the characters in the x-axis label as described in the section "Format Axis Label." The resulting final graph will then look like Figure A2.21.

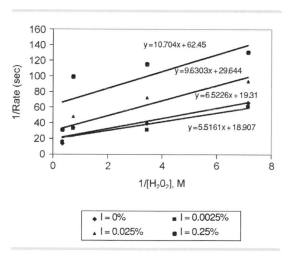

Figure A2.21 Lineweaver-Burk plot showing the effect of inhibitor concentration on rate of enzyme activity with different substrate concentrations

Figures as an Appendix to the Lab Report

Whenever possible, figures should be inserted into the document as close to their textual reference as possible. In some instances, however, you may be required to attach the figures on separate pages at the end of the report. There are two acceptable ways to make the figure caption in that case:

1. Follow the instructions in the sections "Entering Data in Spreadsheet" and "Using Chart Wizard to Plot the Data." In Chart Wizard Step 3 of 4, leave Chart Title blank.
   a. After the figure is done, single-click the area outside of the axes, but inside the frame (Chart Area, not Plot Area).
   b. From the menu bar, select File | Print Preview | Set-up.
   c. Select the Header/Footer tab, and click the Custom Footer box. In the Left section of the footer area (max. 255 characters), type the word *Figure* followed by one space and then the appropriate number, followed by two spaces and then the title. Select the box to change the font (Arial, regular, 12 pt is acceptable). Select O.K. to close all the set-up windows. The figure caption should be displayed in the proper position in the Print Preview window.

2. Follow the instructions in the sections "Entering Data in Spreadsheet" and "Using Chart Wizard to Plot the Data." In Chart Wizard Step 3 of 4, instead of leaving Chart Title blank, type the word *Figure* followed by one space and then the appropriate number, followed by two spaces and then the title.
   a. Note that Excel places the figure caption above the figure. After the figure is done, select the chart title text box and drag it below the graph. This will necessitate selecting the graph and moving it upwards to prevent crowding.
   b. Single-click the area outside of the axes, but inside the frame (Chart Area, not Plot Area).
   c. From the menu bar, select File | Print Preview. This option will allow you to see the figure exactly as it will be printed. Make sure the Header/Footer is set to None. Check this as follows:
      - From the menu bar in Print Preview, select Set-up. You will find four tabs under this option: Page, Margins, Header/Footer, Chart.
      - Select Header/Footer. If the default is None, you are all set.

Importing Tables into Microsoft Word

Tables are handy for recording large amounts of data. Raw data should not, however, be included in your lab report or scientific paper. Furthermore, do not include both a table and a figure when they contain the same data.

Tables are preferable to figures in the following situations:

- To show precise numeric information rather than just the trend (as conveyed by a figure)
- To summarize information
- To describe information that is too complex to be shown in any other form.

Because the Excel worksheet is really just one huge table, it may be easier to enter data in Excel, and then import the table into Word. Follow these steps:

1. Construct a well-organized table in Excel. Columns and their headings usually display the dependent variable, so that like data are compared in the columns, rather than across rows. If the variables presented have units, place the units in parentheses after the title in the column heading.

2. Use the mouse to highlight the data you wish to copy.

3. From the Standard toolbar, select the Copy icon (or from the menu bar, select Edit | Copy).

4. Position the cursor in the Word document at the location where you wish to insert the table.

5. From the Standard toolbar, select the Paste icon (or from the menu bar, select Edit | Paste).

6. Once the table is in the Word document, you can change the table format by selecting Table from the menu bar. Click Hide Gridlines to display or hide the gridlines. With the table selected first, select Table | Table AutoFormat, and then choose an appropriate format from the extensive list. Remember that most scientific journals publish articles in black and white, so avoid formats that use color.

7. Formal tables are always numbered and titled **above** the data. The numbers and titles may be centered or placed flush on the left margin of the report (acceptable table caption formats are similar to figure caption formats. See Item 7 under "Importing Graphs into Microsoft Word"). Arabic numbers are used, and the tables are numbered consecutively in the

order they are discussed in the text. Titles are always single spaced.

8. Table titles should be informative and consist of a precise noun phrase.

9. Tables should not be split across pages unless the table is too large to fit on one page. If it is not possible to fit a table in the remaining space at the bottom of a page, insert a page break and start the table on the next page.

PEER RESPONSE FORMS

DRAFT SELF-ASSESSMENT

Writer's name

Title of paper

1. What are the goals and objectives of this paper? Do I make these objectives clear to the reader?

2. What questions and concerns do I have with this draft? What are some ways I can begin to address these concerns?

3. What parts would I like feedback on (not sure if my meaning is clear, not sure if I understood this concept correctly, and other issues)?

4. What do I like about this paper? What are its strengths and what has gone well?

PEER REVIEW GUIDE

Writer's name

Title of paper

Peer evaluator's name

1. Do I know what the writer is trying to accomplish with this paper? Is the purpose clear?

2. What questions or concerns do I have about this paper? Are there sections that are difficult to follow? Is the organization, content, flow, and level appropriate for the intended audience?

3. What suggestions can I offer the writer to help him/her clarify the intended meaning?

4. What do I like about the paper? What are its strengths?

SENTENCES REQUIRING REVISION

Rule #1 THE AUDIENCE FOR YOUR PAPER IS YOUR PEERS; WRITE AS THOUGHT YOUR PEERS ARE FELLOW SCIENTISTS, NOT STUDENTS IN A CLASSROOM SITUATION.

Rewrite the following sentences for your audience.

1. The purpose of this experiment is to become acquainted with new lab techniques such as serial dilutions, protein analysis, and use of a spectrophotometer.

2. The purpose of this experiment is to teach the student how to determine the protein concentration in an unknown sample.

3. By using the formulas we learned in lab, if we have a protein concentration of 33.75 mg/mL, 10 mL of the sample would contain 337.5 mg of protein.

4. As can be expected in any undergraduate laboratory, many routine technical errors are bound to occur.

5. It is a very common error for students not to adhere exactly to the recommended incubation time.

6. Since this technique is fairly new to first-year students, mistakes could have been made.

Rule #2 ELIMINATE UNNECESSARY INTRODUCTIONS. GET RIGHT TO THE POINT.

Explain why the following sentences, or parts of the sentence, are unnecessary. If you have some experience with the topic, propose a more direct alternative. In some cases, it may be possible to combine sentences. In other cases, deleting a sentence is the best choice.

1.   It is significant to note that the solutions with less inhibitor had higher reaction rates.

2.   This experiment successfully demonstrated the relationship between substrate concentration and enzyme concentration, and its effect on the rate of reaction of an enzyme, catalase.

3.   It was very important to have exact measurements for this lab or else the spectrophotometer would pick up absorbance values, which are either too high or too low.

4.   Assays are commonly used as tests to calculate the activity of an enzyme under these variable conditions, which is necessary in many experiments that deal with enzymes.

5.   The purpose of this experiment will be to achieve a higher level of understanding about the way enzymes work.

6.   Enzymes are found in all biological systems and are composed of multiple peptides joined by peptide bonds.

7.   The purpose of this experiment is to understand what enzymes do, how they react, and how changing the concentrations of the reactants can affect the reaction rate.

8.   All of the results can be used to find the optimal conditions for the enzyme to function and lead to a greater understanding of how an enzyme works.

9.   Figure 1 shows the results for the different enzyme concentrations and the corresponding enzyme activity rates using a 3% hydrogen peroxide solution. Figure 2 shows the Michaelis-Menten graph for the different hydrogen peroxide concentrations and corresponding enzyme activity rates using a 100 units/mL enzyme solution. Figure 3 shows the Lineweaver-Burk graph for the 1/different hydrogen peroxide concentrations and 1/corresponding enzyme activity rates using a 100 units/mL enzyme solution. Figure 4 shows results for different hydrogen peroxide concentrations and the corresponding enzyme activity rates using the optimum temperature and 100 units/mL enzyme solution.

10.   Enzymes function most efficiently under certain conditions, which suggests in what type of environment the cell prefers to grow.

11.   Quantitative analysis of enzymes is a widely explored area.

12.   Displayed in the preceding pages are several tables and figures, which help to better visualize enzymatic reactions.

13. For our personal experiment we tried to find the optimal pH and temperature for the enzyme catalase. According to our findings both pH and temperature have an effect on enzymatic activity.

14. In every biological system that has ever walked, crawled, or existed on earth, the cells within the organism must have ways to deal with metabolic waste. One of the most toxic substances to living organisms is a free radical. Many cells use enzymes to breakdown the harmful waste into something benign to the cell to deal with these types of waste products.

15. Bovine gamma globulin (BGG) was used in the Bradford assay to interact with Coomassie Plus reagent to conduct a quantitative protein analysis and to create a standard curve.

16. The purpose of the experiment was to employ both Biuret and Bradford assays in order to obtain a quantitative protein analysis of an unknown sample.

17. If we had 75% transmittance, this means that not all of the light had been sent through the sample. Had this been a mistake, the final readings would have been off.

18. After 10 minutes had passed, the tubes were taken to a spectrophotometer. The machine was zeroed with the blank at a wavelength set at 550 nm before any readings were taken.

19. Protein concentration of the unknown was determined by taking absorbency readings of known concentration samples and graphing the data.

20. I think the Biuret and Bradford assays are good ones to use, but I couldn't find any references comparing those two assays.

21. Both the Biuret and Bradford methods require preparation for quantitative protein analysis of an unknown.

22. Like in all experiments, this one required a lot of careful measurements.

23. By preparing unique solutions for both the Bradford and Biuret methods, and through the use of a spectrophotometer, the amount of protein in an unknown substance was determined.

24. The protein concentration of a sample can be used for many things and in many scientific applications.

25. This is effective because it is quick and easy.

26. Because diluted solutions were needed, calculations were done and nine test tubes were set up.

27. During the experiment, some interesting observations were seen and then further tested.

Rule #3 USE PUNCTUATION CORRECTLY.

Correct the punctuation usage in the following sentences, rewriting the sentences to improve clarity where necessary.

1. It is known that at certain levels high temperatures will in fact denature the enzyme destroying many of the intermolecular bonds that hold its conformation.

2. The active oxygen derivatives can inhibit the cells function or even kill it.

3. In this particular case study we tested the effects of substrate, and enzyme concentration and also pH, and temperature on the effects of enzymatic reactions. The results concluded that higher concentrations of substrate, and enzyme lead to increased enzymatic rate.

4. The reaction occurred at its quickest rate when salt was not present; meaning that the optimal concentration of salt was 0.0 M.

5. Catalase is an enzyme that breaks down toxic hydrogen peroxide, into water, and oxygen.

6. The results suggested that, light is required for photosynthesis.

7. Other dilutions of unknown had values, which were out of the sensitivity range.

8. Proteins are, "complex macromolecules… (that) comprise in essence the working machinery of life," (Haschemeyer, 1973).

9. "In all areas of biological and medical research today there is increasingly a need for knowledge about proteins," (Haschemeyer, 1973).

10. It was also concluded that, there is a general agreement that the Bradford assay gives lower protein values in a variety of organisms.

11. In this experiment, catalase is obtained from a particular prokaryote; potatoes.

12. Five different enzyme solutions were prepared (20 ml's each) at the following concentrations (units/ml): 100, 75, 50, 25, and 0.

Rule #4 ELIMINATE WORDINESS.

Rewrite the following sentences to eliminate redundant words, empty phrases, and needlessly complex constructions. In some sentences, words are underlined to give you a hint. It may be appropriate to combine or delete sentences in some of the examples.

1. This can be attributed to the fact that perhaps the trendline had too many outlying values that led to an inaccurate slope.

2. Another error was that the trendlines on the inhibitor concentration graph didn't intersect exactly on the y-axis as should have happened.

3. It is known that at certain levels high temperatures will in fact denature the enzyme destroying many of the intermolecular bonds that hold its conformation.

4. Another observation that we performed was the rate of enzyme reaction in different concentrations of enzyme.

5. All of the results can be used to find the optimal conditions for the enzyme to function and lead to a greater understanding of how an enzyme works.

6. In this experiment, the catalase of a potato was extracted and tested for reaction rate under different conditions.

7. One of the conditions that the enzymes are sensitive to is salt concentration.

8. The reason for this is because of its proclivity for a low salt concentration than other enzymes.

9. The purpose of the experiment was to employ both Biuret and Bradford assays in order to obtain a quantitative protein analysis of an unknown sample.

10. From the standard curve generated, the unknown protein concentrations were then compared to the known ones in order to get an idea of what they were.

11. These assays alone cannot tell what the protein concentration of a substance is.

12. Quantitative analysis of proteins is a widely explored area. There are multiple forms of methods for the quantitative analysis, too. The two methods used by our class were the Biuret assay and the Bradford assay.

13. For the Bradford and Biuret assays you can see that the classes results have almost no correlation. We cannot make

a conclusion on the statement made by Berges and colleagues (1992) that the Bradford assay gives us a generally lower protein concentration. We had a lower Bradford concentration for egg yolks, but a higher protein concentration for egg whites. Even though most of our class results seem to be useless, they aren't. We can make conclusions on why parts of our experiment went wrong. Since it was our first experiment of the year, I'm sure everyone was a little bit sloppy and out of practice. Since we were using such small amounts of solutions, even the slightest pipette error will greatly throw off our results. There are many other confounding variables that could have affected our experiment. Since there were multiple experimenters, every ones procedure was a little bit different. A dirty test tube could have led to an impure spectrophotometer reading. The amount of time that the color reagent sat in the protein mixture after being vortexed, could also effect its pigmentation.

14. For the next time I would do this experiment I would be sure to get very accurate pipette measurements. I will also be very careful about the spectrophotometer reading. Keeping the spectrophotometer zeroed is very important in getting an accurate reading. Like in all experiments, this one required a lot of careful measurements.

15. The majority of the error in the results came from inexperience with regard to the new equipment that the groups used.

16. As a result of the logarithmic nature of absorbance to transmittance, there is a reverse in direction for the relationship between transmittance and protein concentration. In other words, the more protein there is in a sample, the less amount of light transmittance will occur through the spectrophotometer to the photocell.

17. BSA is highly reactive in the Bradford assay, thereby effecting underestimation of protein concentration.

18. Visibility of the protein concentration is the result of the Coomassie Brilliant Blue dye being bound to certain amino acids.

19. It is significant to note that the solutions with less inhibitor had higher reaction rates.

20. This experiment successfully demonstrated the relationship between substrate concentration and enzyme concentration, and its effect on the rate of reaction of an enzyme, catalase.

21. It was very important to have exact measurements for this lab or else the spectrophotometer would pick up absorbance values, which are either too high or too low.

22. The second portion of our study involved the testing of the effects of temperature and pH on enzyme activity.

23. The purpose of this experiment was the examination of the effect of different growth conditions on the turning on of the lac operon in *E. coli.*

24. The analyses were done on the recombinant DNA to determine which piece of foreign DNA was inserted into the vector.

25. The review article by Lumsden (2001) presents a summary of the recent literature on circadian rhythms in plants.

26. This data is heavily supported in many different literary sources.

27. Another observation that we performed was the rate of enzyme reaction in different concentrations of enzyme.

28. The word *than* is an expression for comparison of two things.

29. Eight beakers were labeled with the following concentrations of hydrogen peroxide and those solutions were created and placed in the appropriate beaker: 0, 0.1, 0.2, 0.5, 0.8, 1.0, 5.0, and 10.0%.

30. This lab will also serve to further explain the effects of temperature as an inhibitor, as well as be able to determine an optimal enzyme activity temperature.

31. One of the focuses of this experiment is the effect that different temperatures have upon enzyme activity.

32. The findings show that the class average for the Biuret method was 11.12 mg/mL and the Bradford was 14.64 mg/mL.

33. From the Biuret standard curve, the unknown protein concentrations were compared to the known ones in order to get an idea of what they were.

34. These assays alone cannot tell what the protein concentration of a substance is.

35. A Berges, Fisher, and Harrison experiment was conducted in 1992 comparing the Lowry, Bradford, and Smith assays on proteins isolated from marine diatoms. They stated, "It remains unclear which spectrophotometric assay is most accurate, but the Bradford assay is faster and simpler, and is

less likely to be affected by a nonprotein compound found in marine phytoplankton."

36. Disks of filter paper were <u>obtained and placed</u> into the respective enzyme solution.

37. <u>Five beakers were labeled and filled</u> with the following concentrations of ammonium chloride salt: 4, 2, 1, 0.5, 0.1, and 0 M.

Rule #5 ELIMINATE AMBIGUITY.

Rewrite the following sentences to eliminate ambiguity. Avoid vague use of *this, that, which,* and *it* in reference to topics mentioned previously.

1. The activity of enzymes depends heavily upon the conditions in which it is working.

2. The first test was for different concentrations of the catalyst, which had a positive linear relationship, in which the higher the concentration, the faster the enzymes reacted with the hydrogen peroxide. The second test was testing it for different concentrations of hydrogen peroxide.

3. Hotter temperatures cause more contact between the substrate and the enzyme, which resulted in higher enzymatic activity.

4. The standard curve that can be obtained from this graph has an equation for a trendline, that is needed to find the concentration of the unknown samples.

5. This means that the salt changes the conformation of the enzyme as time passes, which makes it less reactive with the substrate.

6. The enzyme was prepared by separating catalase from potato tissue. Once extracted, the disks of filter paper were dipped into the solution of catalase and dropped into the substrate, hydrogen peroxide.

7. The zero substrate concentration had no reaction rate.

8. To prevent the buildup of active oxygen derivatives in its tissue, the potato produces catalase to increase the rate of reaction so it can get rid of the hydrogen peroxide.

Rule #6 USE CONNECTING WORDS AND REPETITION TO IMPROVE FLOW.

Rewrite the following passages to improve flow. Combine sentences, eliminate wordiness, and use appropriate punctuation to avoid confusion.

1. Catalase is an enzyme found in many organisms. We used catalase extracted from potatoes and carrots in our experiment. Catalase is an enzyme that breaks down toxic hydrogen peroxide, into water, and oxygen. Like all enzymes, catalase has a certain range of temperature and pH, for optimal reaction rate. Catalase's optimum pH level is 6, and its optimum temperature level is 30°C, (Min and Mistry, 1992). This data is heavily supported in many different literary sources. Our experiment had a bit different results however. We concluded that catalase has optimal reaction rate at pH 7, and 50°C.

2. To prevent the buildup of active oxygen derivatives in its tissue, the potato creates catalase to increase the rate of reaction so it can get rid of the hydrogen peroxide (Xiaozhong, 2000). The active oxygen derivatives can inhibit the cells function or even kill it. Low or high temperature can lower the rate at which the catalase can react with the hydrogen peroxide. This can lead to a buildup of the toxin. In optimum conditions, the enzyme functions at a rate that will prevent any substantial buildup of the toxin (Xiaozhong, 2000). Heat can lead to the denaturization of the enzymes (Anonymous, 2000). This hydrogen bond breakage results in enzymes that have lost the active sites where the reaction between the enzyme and the substrate occur. The enzyme lowers the activation energy needed for the reaction from hydrogen peroxide to oxygen and water (Craig, 1998).

3. As light passes through the cuvette in the spectrophotometer, the amount of light absorbed can be portrayed in a linear equation. The unknown protein sample's values for absorbance can then be plugged into the equation to determine the protein concentration. The 1:15 dilution of the unknown produced an absorbance value that fell between the sensitivity range (0.2–1.4 mg/mL) and could be substituted into the obtained equation. For example, a 75% transmittance translates that 25% of the light was absorbed which calculates to an absorbance of 0.125.

4. In every biological system that has ever walked, crawled, or existed on earth, the cells within the organism must have ways to deal with metabolic waste. One of the most toxic substances to living organisms is a free radical. Many cells use enzymes to breakdown the harmful waste into something benign to the cell to deal with these types of waste products.

5. After 10 minutes had passed, the tubes were taken to a spectrophotometer with a wavelength set at 550 nm. The machine was zeroed with the blank before any readings were taken.

6. These assays alone cannot tell what the protein concentration of a substance is. Rather, they are used in conjunction with the spectrophotometer, which is a technique that measures the amount of light given off when a wavelength passes through a sample.

7. Errors could have occurred during the experiment when the solutions were being prepared with the pipettes and when the absorbance of the solutions was measured in the spectrophotometer. The majority of the error in the results came from inexperience with regard to the new equipment that the groups used. There was also some confusion in the reporting of the data. If a group put their concentration in the wrong columns, this would greatly skew the overall conclusions.

8. Quantitative analysis of enzymes is a widely explored area. There are multiple forms of methods for the quantitative analysis, too. The two methods used by our class were the Biuret assay and the Bradford assay.

9. For the Bradford and Biuret assays you can see that the classes results have almost no correlation. We cannot make a conclusion on the statement made by Berges and colleagues (1992) that the Bradford assay gives us a generally lower protein concentration. We had a lower Bradford concentration for egg yolks, but a higher protein concentration for egg whites. Even though most of our class results seem to be useless, they aren't. We can make conclusions on why parts of our experiment went wrong. Since it was our first experiment of the year, I'm sure everyone was a little bit sloppy and out of practice. Since we were using such small amounts of solutions, even the slightest pipette error will greatly throw off our results. There are many other confounding variables that could have affected our experiment. Since there were multiple experimenters, every ones procedure was a little bit different. A dirty test tube could have led to an impure spectrophotometer reading. The amount of time that the color reagent sat in the protein mixture after being vortexed, could also effect its pigmentation.

10. For the next time I would do this experiment I would be sure to get very accurate pipette measurements. I will also be very

careful about the spectrophotometer reading. Keeping the spectrophotometer zeroed is very important in getting an accurate reading. Like in all experiments, this one required a lot of careful measurements.

Rule #7 MAKE SUBJECTS AND VERBS AGREE.

Underline the subject of each of the following sentences. Correct the associated verb to make it agree with the subject.

1. The activity of enzymes depend heavily upon the conditions in which it is working.

2. This data is heavily supported in the literature.

3 One of the enzymes that are extensively studied is catalase, which is found in peroxisomes of the cell.

4. The collected data was then plotted, x-axis being protein concentration and y-axis as absorbance.

5. One drawback of these methods are that they require a long incubation period.

6. There is a linear relationship between the protein concentration and their respective absorbance values. The higher the concentration, the higher the absorbance value.

7. The reaction time for varying solutions of hydroxylamine in varying solutions of hydrogen peroxide are given in Table 1.

8. The only valid data for the Biuret method was that which fell within the sensitivity range of 1 to 10 mg/mL.

9. Approximately 5 drops of inhibitor was added to each substrate concentration.

10. Approximately 5 mL of inhibitor were added to each substrate concentration.

Rule #8 WRITE IN COMPLETE SENTENCES.

Some of the following examples contain sentence fragments. Revise each passage by combining sentences or deleting the subordinate word(s) to make a complete sentence.

1. It is known that at certain levels high temperatures will in fact denature the enzyme destroying many of the intermolecular bonds that hold its conformation. Thus changing the effectiveness of the enzyme.

2. The equation of the trendline was used to derive the protein concentration of the various dilutions of egg yolk. Although only those whose protein concentrations fell within the sensitivity range of the assay would be multiplied by the dilution factor to give the unknown concentration value.

3. The reaction occurred at its quickest rate when salt was not present; meaning that the optimal concentration of salt was 0.0 M.

4. In the second experiment, the solution varying salicylic acid concentrations.

Rule #9 REVISE RUN-ON SENTENCES.

Revise the following run-on sentences using one or more strategies:

- Insert a comma and a coordinating conjunction (*and, but, or, nor, for, so,* or *yet*)
- Use a semicolon or possibly a colon
- Make two separate sentences
- Rewrite the sentence

1. An increase in enzyme concentration quickened the reaction rate as did an increase in substrate concentration, so the concentrations of the molecules have an influence on how the enzyme reacts.

2. Note that although the graphs shows the rapid increase of reaction rate at higher levels of substrate concentration, it is not readily apparent as usually is on a Michaelis-Menten graph—and therefore must be extrapolated—that that rate will start to level off as it approaches the V_{max}, or the point at which the enzymes present in the solution are reacting as fast as possible, and there is an excess of substrate.

3. For example, using Figure 2, if one were to wonder what would happen if the substrate concentration were increased to 20% hydrogen peroxide, (s)he could look at the Michaelis-Menten plot and hypothesize that around that concentration of substrate, the rate of reaction would be at its V_{max}, because it would seem that the catalase would be reacting as fast as possible, and there is a much larger amount of substrate present than can be reacted with at a given time.

4. Competitive inhibitors actually bond to the active site of the enzyme and physically block the substrate from binding, but the noncompetitive inhibitors merely bond to another part of the enzyme, altering the shape of the active site, making it impossible for the substrate to bond to the enzyme.

5. Enzymes are large proteins that are made up of hundreds of amino acids that often contain a prosthetic group that is important in the chemical reactions that are important in living cells.

6. Catalase is an oxidizing enzyme present in the peroxisomes of nearly all aerobic cells, serves to protect the cell from the toxic effects of hydrogen peroxide by catalyzing its decomposition into molecular oxygen and water without the production of free radicals.

7. The readings from the spectrophotometer should show a linear correlation of protein intensity and absorbency, this is the Beer-Lambert Law, which relates absorbency to the path length of light along with molar concentration of a solute and the molar coefficient.

8. The proteins are "tagged" by certain chemicals that react in a predictable and visible manner based on the intrinsic properties of the proteins, which allow them to be viewed and analyzed by such instruments as the spectrophotometer.

9. Visibility of the protein concentration is the result of the Coomassie Brilliant Blue dye being bound to the proteins on account of the dye's sulfonic acid groups' electrostatic attraction, which also results in an absorbance peak at 595 nm in the spectrophotometer.

10. To perform the Bradford method, the standard that was used was bovine gamma globulin or BGG, along with TBS as a solvent, Coomassie Brilliant Blue to find to the proteins, and the sample of unknown protein concentration, the egg yolk.

11. Finally, a graph of the absorbencies versus the protein concentration was plotted in Excel to determine a trendline, whose equation would allow the measured absorbencies from the unknown dilutions to be substituted for y so as to derive the actual protein concentration.

12. The Lineweaver-Burk graph is better because it presents the data in linear form, which more easily allows an equation for the line to be derived, and there by allowing the K_m and V_{max} to be calculated more easily.

Rule #10 BEWARE OF MISUSED WORD PAIRS. USE THE RIGHT WORD FOR THE SITUATION.

1. Catalase was also (affected, effected) by different pHs.

2. The reason for this effect was probably the result of alterations in the (binding, bonding) (confirmation, conformation) due to the different ions present in the solution.

3. Figure 1 shows the class averages for the reaction rates of catalase in the presence of (various, varying) amounts of the enzyme.

4. These factors include temperature, pH, and inhibitors that (bind, bond) to an enzyme preventing the reaction from taking place.

5. Competitive inhibitors actually (bind, bond) to the active site of the enzyme and physically block the substrate from (binding, bonding), but the noncompetitive inhibitors merely (bind, bond) to another part of the enzyme, altering the shape of the active site, making it impossible for the substrate to (bind, bond) to the enzyme.

6. To prevent the buildup of active oxygen derivatives in its tissue, the potato (creates, prepares, produces) catalase to increase the rate of reaction so it can get rid of the hydrogen peroxide.

7. The longer the solution was made and sat inactive, the more that sodium chloride had an (affect, effect) on the active site, slowing down the reaction rate.

8. In this study, the (affect, effect) of the salt NH_4Cl is shown. It is expected that catalase will be uninhibited by small salt concentrations, and (devastated, _____) by high salt concentrations.

9. This indicates that there is a limit to how quickly catalase (can operate on, _____) hydrogen peroxide.

10. This means that the salt changes the (confirmation, conformation) of the enzyme as time passes, which makes it less reactive with the substrate.

11. The results conclude that increasing protein concentration increases (absorbance, absorbency).

12. In this experiment, we used the Biuret and Bradford methods to determine the amount of protein in an unknown by measuring the (absorbance, absorbency).

13. This brand of paper towel is cost effective because of (its, it's) superior (absorbance, absorbency). You can clean up a spill with fewer paper towels.

14. New parents prefer this brand of diaper because of (its, it's) superior (absorbance, absorbency). The material soaks up more liquid (than, then) other brands of diaper.

15. Spectrophotometric readings measure the intensity of the color and enable the researcher to (decipher, determine) the protein concentration of a sample.

16. There is a linear relationship between protein concentrations and their respective (absorbance, observance) values. The higher the concentration, the higher the (absorbance, observance) value.

17. The Bradford method is preferred because of (its, it's) greater sensitivity.

18. When you have questions, (its, it's) prudent to contact your instructor well before the exam.

19. Always remember that (its, it's) is a contraction of *it is.*

20. Always remember that (its, it's) is the possessive form of *it.* *(Its, It's)* means belonging to "it."

21. The Spec 20 is not accurate when (its, it's) absorbance readings exceed 1.0.

22. The catalase filter-disk assay is used in many introductory biology labs because (its, it's) fast and requires no expensive supplies.

23. I like to use Science Citation Index to find references. (Its, It's) format is user friendly and (its, it's) the only database that allows you to search both forward and backward in the literature.

24. Each group of students would not have time to do the lac operon lab in (its, it's) entirety, because there are too many sample flasks to monitor.

25. In the scientific method, first ask a question, and (than, then) turn your question into a hypothesis.

26. When we extracted catalase from potatoes, first we chopped up 50 g of potato in a blender, and (than, then) we filtered the homogenate.

27. When you compare two items use (than, then) not (than, then).

28. The experiment took longer (than, then) I thought it would.

29. The Biuret method typically gives higher protein concentrations (than, then) the Bradford method.

30. Ethylene glycol is thought to be a safer preservative (than, then) formalin.

31. Bucknell's cross country program has traditionally been stronger (than, then) that of other Patriot League schools.

32. If the wrestling team wins the league championship, (than, then) it will advance to the ECACs.

33. My grade is higher (than, then) the class average.

34. Our class results concluded that egg yolks contain a higher concentration of protein (than, then) do egg whites.

35. This means that, with time, the salt changes the (confirmation, conformation) of the enzyme.

36. Five drops of each (various, varying) dilution of salicylic acid was added to 1 mL of catalase.

37. The reaction times for (various, varying) solutions of hydroxylamine in (various, varying) solutions of hydrogen peroxide are given in Table 1.

38. The differing protein concentrations determined by the Biuret and Bradford assays might be explained by the (various, varying) sensitivity to the protein itself as well as interference by other substances in the sample.

39. The reaction rate will continue to increase until the (amount, number) of active sites exceeds the (amount, number) of substrate molecules.

40. The (amount, number) of time required for the disks to float was measured.

41. The greater (amount, number) of catalase molecules that can bind with the substrate molecules explains this increase in reaction rate.

42. The following hydrogen peroxide solutions were (created, prepared, produced): 0, 0.1, 0.2, 0.5, 0.8, 1.0, 5.0, and 10.0%.

43. Each solution was (created, prepared, produced) just before it was tested.

44. The catalase-soaked disk was dropped into (various, varying) concentrations of hydrogen peroxide.

45. The same amount of each (various, varying) concentration of BSA was used for the absorbance measurement.

46. If hydrogen peroxide concentration was high, and salicylic acid was low, (than, then) the reaction would proceed (at a high rate, quickly).

47. The pH (affects, effects) the substrate and enzyme activity due to hydrogen dissociation.

Rule #11 USE DISCRETION WHEN USING SPELL CHECKERS.

Rewrite the following passage, changing spelling to make the sentences make sense.

Wrest a Spell

Eye halve a spelling chequer
It came with my pea sea
It plainly marques four my revue
Miss steaks eye kin knot sea.

Eye strike a key and type a word
And weight four it two say
Weather eye am wrong oar write
It shows me strait a weigh.

As soon as a mist ache is maid
It nose bee fore two long
And eye can put the error rite
Its rare lea ever wrong.

Eye have run this poem threw it
I am shore your pleased two no
Its letter perfect awl the weigh
My chequer tolled me sew.

– Sauce unknown

BIBLIOGRAPHY

Bregman A. 1996. Laboratory investigations in cell and molecular biology. Rev. 3rd ed. New York: John Wiley & Sons. 336 p.

Calabria J, Burke D. 1997. Microsoft Word 97 exam guide. Indianapolis: QUE Corp.

Corrections to Publications: Corrections to Scientific Style and Format (6th ed.) Updated June 2000 Available at <http://www.councilscienceeditors.org/pubs_corrections .shtml> (Retrieved on June 13, 2001).

Glenn W. 2000. Word 2000 in a nutshell. Sebastopol, CA: O'Reilly & Associates, Inc. 491 p.

Hacker D. 1997. A pocket style manual. 2nd ed. Boston: Bedford Books. 185 p.

Hailman JP, Strier KB. 1997. Planning, proposing, and presenting science effectively. Cambridge: Cambridge University Press. 150 p.

Harnack A, Kleppinger E. 2001. Online! A reference guide to using internet sources. Boston: Bedford/St. Martins. 260 p.

Information Services and Resources, Bucknell University. 2001. Evaluating Web Resources. Available at <http://www.isr.bucknell.edu/research /evaluating.html> (Retrieved on October 5, 2001).

Jones A, Reed R, Weyers J. 1994. Practical skills in biology. London: Longman Scientific & Technical. 292 p.

Light RJ. 2001. Making the most of college: Students speak their minds. Cambridge, MA: Harvard University Press. 242 p.

Lunsford A. 2002. The everyday writer. 2nd ed. Boston: Bedford/St. Martin's. 534 p.

Lunsford A, Connors R. 1999. Easy writer: A pocket guide. Boston: Bedford/St. Martin's. 266 p.

Lunsford A, Connors R. 1995. The new St. Martin's handbook. 3rd ed. Boston: Bedford/St. Martin's. 797 p.

McMillan VE. 1997. Writing papers in the biological sciences. 2nd ed. Boston: Bedford Books. 197 p.

Moore DS. 2000. The basic practice of statistics. 2nd ed. New York: W. H. Freeman and Company. 619 p.

Palmer-Stone D. Last revised July 31, 2001. How to Read University Texts or Journal Articles. Counselling Services, University of Victoria, Victoria, BC, Canada. Available at <http://www.coun.uvic.ca/learn/program/hndouts /Readtxt.html> (Retrieved on October 2, 2001).

Pechenik JA. 1997. A short guide to writing about biology. 3rd ed. New York: Longman. 318 p.

Peterson SM. 1999. Council of Biology Editors guidelines No. 2: Editing science graphs. Reston, VA: Council of Biology Editors. 34 p.

Peterson SM, Eastwood S. 1999. Council of Biology Editors guidelines No. 1: Posters and poster sessions. Reston, VA: Council of Biology Editors. 15 p.

Samuels ML, Witmer JA. 1999. Statistics for the life sciences. 2nd ed. Upper Saddle River, NJ: Prentice Hall. 683 p.

Style Manual Committee, Council of Biology Editors. 1994. Scientific style and format: The CBE manual for authors, editors, and publishers. 6th ed. Cambridge, MA: Cambridge University Press. 825 p.

The University of Reading, IT Services. 1998. Equations and fields in Microsoft Word 97. Available at <http://www.rdg.ac.uk/ITS /Topic/WordProc/WoPW7equ01> (accessed on August 24, 2001).

Utts JM, Heckard RF. 2002. Mind on statistics. Pacific Grove, CA: Duxbury Press. 592 p.

VanAlstyne JS. 1986. Professional and technical writing strategies. Upper Saddle River, NJ: Prentice Hall, Inc. 433 p.

Vodopich D, Moore R. 1996. Biology laboratory manual to accompany biology. 4th ed. Boston: WCB/McGraw-Hill. 510 p.

Watson W. 1999. Equation editor tutorial. Mechanical Engineering Department, Texas Tech University, Engineering Analysis I (ME1305). Available at <http://www.osci.ttu.edu/me1305 /word/tutorial.html> (accessed on August 24, 2001).

Woolsey JD. 1989. Combating poster fatigue: How to use visual grammar and analysis to effect better visual communications. Trends in Neuroscience 12(9): 325–332

INDEX

Abbreviations, 56
  table of, 58–59
absorbance/absorbency, correct usage,
  70
Abstract, writing for lab report, 40–41
Active voice, 54–55
affect, correct usage, 71
Aligning text, 113–120
  quick reference chart, 112
AltaVista, 14
Ambiguity, 66
amount, correct usage, 71
Annual Reviews, 11
ArticleFinder, 10, 13
Ask Jeeves!, 14
Audience, for scientific papers, 53
AutoCorrect, 138–139
AutoSave, 102

Bar graphs, 6, 146
Basic BIOSIS (FirstSearch), 10, 12
Bibliography, 42. *See* References
bind, correct usage, 71
Biology, writing conventions, 53–55
bond, correct usage, 71
Book, reference format, 45
  article in (reference format), 44
Books, as primary references, 11–12
Bulleted lists, 117–120

CBE manual, *see* Council of Biology
  Editors manual
Chart Wizard, 154

Citation format
  in lab report, 42–43
  for posters, 98
Citation-Sequence system (reference
  format), 20
  examples of, 44–46
Clarity, in scientific writing, 62–67
  ambiguity, 66
  complex sentences, 63
  connecting words, 66–67
  empty phrases, 63–65
  redundancy, 63
Clauses, subordinate, 68
Colon, 61
  run-on sentences and, 69, 70
Comma, 56–57, 60
  run-on sentences and, 69
complementary/complimentary, correct
  usage, 71
Complex sentences, 63
Computers, 29, 31
  See also Microsoft Excel; Microsoft
  Word
Conclusions section
  posters, 97–98
  See also Discussion section.
confirmation/conformation, correct
  usage, 71
Conjunctions, run–on sentences and,
  68–69
Connecting words, 66–67
continual/continuous, correct usage,
  71–72
Control treatment, 4